Barrier Islands

From the Gulf of St. Lawrence

To the Gulf of Mexico

Academic Press Rapid Manuscript Reproduction

The Proceedings of a Coastal Research Symposium held on March 9, 1978 in Boston, Massachusetts for the Geological Society of America, Northeast Section and the Society of Economic Paleontologists and Mineralogists, Eastern Section.

BARRIER ISLANDS

FROM THE GULF OF ST. LAWRENCE
TO THE GULF OF MEXICO

Edited by

Stephen P. Leatherman

National Park Service
Cooperative Research Unit
University of Massachusetts
Amherst, Massachusetts

ACADEMIC PRESS

New York   San Francisco   London   1979

A Subsidiary of Harcourt Brace Jovanovich, Publishers

ACADEMIC PRESS, INC.
111 Fifth Avenue, New York, New York 10003

United Kingdom Edition published by
ACADEMIC PRESS, INC. (LONDON) LTD.
24/28 Oval Road, London NW1 7DX

Library of Congress Cataloging in Publication Data

Coastal Research Symposium, Boston, 1978.
    Barrier islands from the Gulf of St. Lawrence to the
Gulf of Mexico.

    Papers presented at the symposium organized for the
Geological Society of America, Northeast Section and
the Society of Economic Paleontologists and Mineralogists,
Eastern Section, held in Boston on Mar. 9, 1978.
    1.   Barrier islands—Atlantic coast (North America)—
Congresses.   2.   Coast changes—Atlantic coast (North
America)—Congresses.   I.   Leatherman, Stephen P.
II.   Geological Society of America. Northeastern
Section.   III.   Society of Economic Paleontologists and
Mineralogists. Eastern Section.   IV.   Title.
GB473.C62   1978   551.4'2   79-15954
ISBN 0-12-440260-7

PRINTED IN THE UNITED STATES OF AMERICA

79 80 81 82   9 8 7 6 5 4 3 2 1

CONTENTS

CONTRIBUTORS

Numbers in parentheses indicate the pages on which authors' contributions begin.

ELIZABETH A. ALLEN (149), The Shell Oil Company, New Orleans Louisiana

JOHN W. ARMON (65), Department of Geology, Memphis State University Memphis, Tennessee

DANIEL F. BELKNAP (149), Department of Geology, University of Delaware, Newark, Delaware

WILLIAM J. CLEARY (237), Program in Marine Sciences, University of North Carolina, Wilmington, North Carolina

JOHN J. FISHER (127), Department of Geology, University of Rhode Island, Kingston, Rhode Island

PAUL J. GODFREY (99), Department of Botany, University of Massachusetts, Amherst, Massachusetts

SUSAN D. HALSEY[1] (185), Department of Geology, University of Delaware, Newark, Delaware

MILES O. HAYES (1), Department of Geology, University of South Carolina, Columbia, South Carolina

S. DUNCAN HERON, JR. (211), Department of Geology, Duke University, Durham, North Carolina

PAUL E. HOSIER (237), Program in Marine Sciences, University of North Carolina, Wilmington, North Carolina

[1]Present address: New Jersey Department of Environmental Protection, Bureau of Geology & Topography, Trenton, New Jersey

CHACKO J. JOHN (149), Department of Geology, University of Delaware, Newark, Delaware

JOHN C. KRAFT (149), Department of Geology, University of Delaware, Newark, Delaware

STEPHEN P. LEATHERMAN (99), National Park Service, Cooperative Research Unit, University of Massachusetts, Amherst, Massachusetts

EVELYN M. MAURMEYER (149), Department of Geology, University of Delaware, Newark, Delaware

S. B. McCANN (29), Department of Geography, McMaster University, Hamilton, Ontario, Canada

THOMAS F. MOSLOW (211), Department of Geology, University of South Carolina, Columbia, South Carolina

GEORGE F. OERTEL (273), Institute of Oceanography, Old Dominion University, Norfolk, Virginia

ERVIN G. OTVOS, JR. (291), Gulf Coast Research Laboratory, Ocean Springs, Mississippi

PETER S. ROSEN (81), Geological Survey of Canada, Bedford Institute of Oceanography, Dartmouth, Nova Scotia, Canada

ELIZABETH J. SIMPSON (127), Department of Geology, University of Rhode Island, Kingston, Rhode Island

ROBERT ZAREMBA (99), Department of Botany, University of Massachusetts, Amherst, Massachusetts

PREFACE

There has been a quantum increase in research on, and hence understanding of barrier islands during the past decade. Articles by Pierce, Dillon, Scott, and others in 1969–1970 marked the beginning of the great surge in barrier island research. A few excellent studies appeared prior to this time, but most of these early papers were largely descriptive in nature. One impetus behind this recent scientific activity has been the accelerated rate of coastal zone development. Since most sandy beaches being subjected to such pressures along the East and Gulf Coasts of North America are barrier beaches (islands or spits), research emphasis has been focused on these dynamic landforms. The trend toward island development has been countered, in part, by conservation policies, such as the creation of coastal parks. Controversy regarding the developmental potential and management strategies for barrier islands has been most important in setting the stage for this research. It is imperative that the processes that formed and continue to shape these mobile coastal environments be understood so that their potential uses can be fully assessed.

The study of modern sedimentary environments has also been useful in interpretation of the rock record. Analysis of vertical sedimentary sequences of barrier islands and associated environments has found much application in prospecting for petroleum reservoirs in ancient, now-buried barrier islands.

Due to the wide-ranging interest in barrier islands, and to recent completion of a significant number of major studies on this subject, a Coastal Research Symposium was organized for the Geological Society of America, Northeast Section, and the Society of Economic Paleontologists and Mineralogists, Eastern Section, and was held in Boston on March 9, 1978. From this symposium, papers focusing on barrier island dynamics have been selected for presentation in this volume, the first dedicated solely to research on Holocene transgressive barriers. Two papers that were not presented at this symposium have been added in order to provide a more comprehensive and less regionalized scope to this volume.

While there are regressive barriers, such as Kiawah and Galveston Islands, most of the world's barrier islands are presently eroding due to a eustatic rise in sea level. Barrier islands are unique among coastal structures in their ability to migrate landward, generally up a gradually sloping coastal plain, while mainland shores must continue to erode in place. It is clear that there are three processes responsible for landward barrier island migration: inlet dynamics, overwash processes, and dune migration (aeolian activity). Although these processes are conceptually well-defined, assessment of the relative importance of each process is fundamental to an understanding of barrier dynamics and development of predictive models. Without this type of quantification and comprehensive interpretation of the geologic past, it is impossible to make accurate forecasts of barrier island evolution. Hence, it is pure speculation to anticipate the environmental effects of various programs of barrier manipulation, such as construction of barrier dunes and stabilization of inlets, without these detailed studies. These research papers have been assembled to present more specific information on barrier dynamics and to provide generalized models that can be applied to petroleum exploration and barrier island management.

The firstchapter, by Miles O. Hayes, summarizes the extensive barrier studies conducted by University of South Carolina researchers worldwide and serves as an excellent introduction to this volume, providing an overview of barrier island characteristics. Examination of embayments, including the German and Georgia Bights, showed a continuum of change in shoreline morphology in response to changes in tidal range and wave energy. Hayes' delineation of the interaction of these two variables in the determination of hydraulic regime represents a significant contribution to our understanding of systematic barrier island variation.

S. Brian McCann's chapter presents a comprehensive review of the Canadian literature and provides the general setting for this and the following two papers. The Gulf of St. Lawrence is characterized by a low-tidal range, locally generated waves, and winter sea ice. His paper closely examines the environmental factors that influence variations in barrier island morphology. McCann recognized that while tidal inlets play a major role in landward transfers of sediment along the transgressive barriers, there exists a large variability in the importance of various processes and, hence, in barrier response within a small enclosed sea.

John W. Armon provides the quantification of these landward transfers for the Malpeque barrier in his chapter. Over 90% of the landward sediment transport occurred in the vicinity of existing or closed inlets during the period of historical photographic record. These quantitative estimates were obtained through a combination of air photo studies and field survey measurements. Storm overwash and wind transport result in significant volumetric transfers at sites of former inlets. A well-vegetated continuous dune largely prevents landward sediment movement, except at inlet-associated areas. Tidal inlets, through development of their large flood tidal deltas, produce the wider zones along the islands and were found to play the dominant role in maintaining the barrier during transgression.

In a contrasting barrier environment, characterized by low dunes and washovers, Peter S. Rosen conducted an intensive short-term field investigation of aeolian transport. For his study, the Tabusintac barrier in the Gulf of St. Lawrence was chosen. Volumetric rates of sand transport by wind were determined by using sand traps and field surveys to monitor surface elevation changes. Washovers served as corridors for aeolian transport, while little sand was found to move across the crest of erosional dunes. Rosen showed that redistribution of sand by wind from washover fans into developing dunes resulted in vertical growth of the barrier.

In the paper by Paul J. Godfrey and associates, a case for the importance of plant types in determining barrier topography is presented. The regional variation in plant species was used to partially explain differences in barrier morphology, comparing Cape Cod, Massachusetts, and Cape Lookout, North Carolina. The two basic plant communities of any barrier—the dune and salt marsh grasses—are dominated by different species or subspecies (varieties) along the East Coast. This chapter introduces botanical measures that can be used or at least should be considered by coastal geologists in their evaluation of a barrier system. It is important to recognize that different ecological mechanisms exist in response to similar geologic processes. Thus, it is not possible to take a model, derived and calibrated for one area, and apply it unilaterally along the East Coast. The findings of this study have clear implications for barrier island management.

Along the Rhode Island coastline John J. Fisher and Elizabeth J. Simpson studied washover and tidal sedimentation rates as environmental factors in development of a transgressive barrier shoreline. A photogrammetric survey of historical shoreline changes was used to document the relative sediment transfers by these two processes. Tidal delta sedimentation accounted for the majority of the landward transport, but overwash was more effective in subaerial deposition. A direct correlation was found between overwash occurrence and beach erosion: overwash is associated with accelerated beach erosion. Along this slowly retreating shoreline, most of the eroded sediment is moved offshore to compensate for sea level rise (in accordance with Bruun Rule), while the remainder is displaced landward as inlet and washover deposits.

The chapter by John C. Kraft and his students is exceptional and represents a landmark contribution. According to Rufus LeBlanc (personal communication, 1979), Kraft's work represents the most thorough and detailed documentation of the vertical sequence of transgressive barriers; more is known about the Delaware barriers than any other such shoreline in the world. This paper is based on extensive subsurface (core) and morphologic data. Interpretation of internal structures and construction of stratigraphic sequences allowed for the formulation of a number of models for barrier transgression. The barriers are moving upward and landward in space and through time. Kraft and others were able to clearly define the interrelationships among factors that result in the different mechanisms for barrier transgression.

Using Kraft's models, Susan D. Halsey showed that the pre-Holocene erosional surface has a strong influence on transgression barriers along developing coasts. Halsey derived a new model to genetically explain the development of barrier systems along the Delmarva Peninsula. Nexus, which is the linking of modern (Holocene) with old topographic features, is actually a model of synthesis. It includes portions of Hoyt's "engulfed beach ridge" and Fisher's "spit segmentation" models, but Nexus is conceptually different since much more emphasis is placed on underlying topography and sediment supply in evolution of the barrier environments.

It is of interest to compare Halsey's chapter with that by Thomas F. Moslow and S. Duncan Heron in terms of age correlations. The latter describes pre-Holocene sediments encountered by coring, but it focuses on the Holocene evolution of Core Banks. The rate of island migration seems to control (or is at least correlated with) the dominant process for migration. During rapid island migration, there is a dominance of overwash compared to the work accomplished by hydraulically active inlets. During the past 4000 years the Holocene record indicates that inlet formation, lateral migration, and closure were the dominant processes for landward migration. Presently, Core Sound is quite shallow, and there are no permanent inlets along this shoreline.

William J. Cleary and Paul E. Hosier applied their geologic and ecologic backgrounds to interpret the environmental evolution of the southern North Carolina coastline. From historical photographic analysis and field surveys, it was possible to define different geomorphic sections along this barrier island chain. A cyclic pattern of recovery following washover was recognized and was found to differ according to grain size distribution of overwash sediments. Cleary and Hosier showed that physiographic and vegetative indicators can be coupled to provide a predictive tool for determination of areas prone to overwash.

The last two chapters of this volume were not presented at the meeting, but are included as important contributions concerning the southeast Atlantic and Gulf Coasts. George F. Oertel studied the Georgia embayment, a mesotidal coast where Holocene barriers have been welded onto Pleistocene features to form the renowned Sea Islands. Where influenced by major rivers, these barriers were separated from the mainland by marsh-filled lagoons and were oriented obliquely to the relict (Pleistocene) shoreline. These latter barriers exhibit a relatively slower rate of landward migration due to the presence of a large sand source. As Hayes has shown in this volume, the wave-dominated margins and tide-dominated center of the Georgia embayment clearly affect the morphology of the barrier islands and their relative stability. Oertel also used the recurved beach ridge/dune patterns adjacent to the inlets to interpret the depositional history of these islands.

In the final chapter Ervin G. Otvos, Jr. used core data, old charts, and historical air photographs to document island migration, drowning, and subsequent redevelopment. Otvos' contention that islands evolved by shoal-bar aggradation has recently been supported by R. A. Davis (AAPG-SEPM Annual Convention, 1979), who has provided actual photographic documentation for such an occurrence along the Gulf Coast of Florida. Otvos showed through comparison of charts that hurricanes could result in the complete destruction of preexisting barriers with poststorm island reemergence in more landward positions. Lateral (westward) island migration is quite rapid along this barrier chain in response to the dominant direction of littoral drift.

In summary, the papers assembled in this book provide a wealth of information on barrier islands from the Gulf of St. Lawrence to the Gulf of Mexico. Clearly, much more work remains to be done in terms of precisely defining the dynamics of individual islands, as well as in formulating conceptual frameworks for the modeling of barrier island migration. These papers represent the most detailed and quantitative work on transgressive barrier islands conducted to date. It is believed that this book will be useful to coastal geomorphologists, petroleum geologists, and barrier island managers and serve to direct future research efforts in this field.

The editor was assisted by Dr. Benno Brenninkmeyer of Boston College and Dr. John Southard of Massachusetts Institute of Technology in the review of submitted abstracts, preparation of the program, and chairing of the Coastal Research Symposium. Dr. John Southard also made a preliminary review of the manuscripts, and his help in this regard is gratefully acknowledged. The camera-ready copy was prepared by Ms. Miriam Leader. The continued support by the National Park Service over the past three years made possible the symposium and the ultimate publication of these proceedings.

MEMORIAL TRIBUTE
to
John H. Hoyt (1928–1970)

John H. Hoyt was a professor of geology at the University of Georgia's Marine Institute at Sapelo Island when he met his untimely death. He was a prolific writer in such fields as nearshore and coastal processes of Holocene and Pleistocene landforms, particularly the genesis and development of barrier islands. Vern Henry, his coauthor of many papers, remarked that he "was just reaching his professional maturity" when he died (Henry, 1974).

In 1977, the idea for a coastal symposium was suggested to be part of the program at the Boston Northeastern Section meeting of the Geological Society of America. It was the tenth anniversary of the publication of Hoyt's (1967) paper "Barrier Island Formation." The symposium, held in 1978, marked what would have been his 50th birthday year. The importance and influence of Hoyt's work, particularly his "Barrier Island Formation" paper, can be seen in the papers included in this symposium volume. Most of the authors first read this paper as undergraduates, under the tutelage of the other authors. We all, in turn, consider it a required paper for our students. Words from the Kaddish are appropriate here: "They still live on earth in the acts . . . they performed and in the hearts of those who cherish their memory." May each of us strive to be as significant a scientist as was John H. Hoyt.

Susan D. Halsey

REFERENCE

Henry, V. J., Jr., 1974, Memorial to John Hagar Hoyt (1928–1970): The Geological Society of America, Memorials, Volume III, pg. 117–122

BARRIER ISLAND MORPHOLOGY
AS A FUNCTION OF TIDAL AND WAVE REGIME

Miles O. Hayes

Department of Geology
University of South Carolina
Columbia, South Carolina

*Barrier islands, which occur primarily on coastal plain
shorelines located on the trailing edges of continents and on
marginal seas, vary in morphology in response to the interaction of tidal range and wave energy effects. Coastal plain
shorelines with medium wave energy (mean wave height (H) =
60-150 cm) exhibit distinct differences in morphology in areas
with different tidal ranges. For example, barrier islands do
not occur on macrotidal coasts (tidal range (T.R.)> 4 m). On
microtidal coasts (T.R. < 2 m), which have the greatest abundance of barrier islands, the barriers are long and linear, with
a predominance of storm washover influences. On mesotidal
coasts (T.R. = 2-4 m), the barriers are short and stunted, with
a characteristic drumstick shape. The large ebb-tidal deltas
that are common on mesotidal coasts of medium wave energy play
an important role in shaping the morphology of adjacent barrier
islands by storing large volumes of sand which becomes available
to the island on occasion and by strongly influencing wave-
refraction patterns.*

*A plot of 21 coastal plain shorelines on a graph of mean
wave height versus mean tidal range allowed further discrimi-
nation of the impact of these two factors on coastal morphology.
In areas of low wave energy (H < 60 cm), smaller tidal ranges
are required to produce tide-dominant morphology than on medium
wave energy coasts. In areas of high wave energy (H > 150 cm),
larger tidal ranges are required.*

*Embayments in coastal plain shorelines exhibit a continuum
of change in shoreline morphology around their margins. Entrances to the embayments are characterized by more wave-
dominant features, with tide-dominant features increasing toward*

1

the heads of the embayments. Examples of these trends are found in the German, Georgia and Western Florida Bights.

Recognition of these responses in barrier island morphology to differences in tidal and wave regime has important implications with regard to engineering procedures applied in the human development of barrier islands, as well as to the interpretation of ancient barrier island sediments in the rock record.

INTRODUCTION

Barrier island shorelines occur primarily along coastal plains located on the trailing edges of continents and on marginal sea coasts (classification scheme of Inman and Nordstrom, 1971). According to a recent study by Glaeser (1978), 76% of the barrier islands in the world are found on coasts of these two types. Exceptions occur in areas where abundant sediment supply, coupled with moderate to high wave energy and small tidal ranges, prevails (for example, the barrier islands of the Copper River delta area, Alaska, which are on a collision coast; Hayes et al. 1976). The worldwide distribution of barrier islands and lagoons (after Gierloff-Emden 1961) is shown in Figure 1. A discussion of the origin of barrier islands is beyond the scope of this paper, which is concerned primarily with differences in barrier island morphology. For details on barrier island origin, refer to Field and Duane (1976), Swift (1975), Wanless (1974) and Hoyt (1967).

Within coastal plain shorelines, barrier islands are restricted to those areas with tidal ranges less than approximately 4 m. This conclusion is based on a study of coastal charts of the world, conducted at the Defense Research Laboratory, University of Texas (in 1963-64; discussed by Hayes 1965 and Hayes and Kana, 1976). The lack of extensive barrier islands in areas with large tides has been pointed out by several other authors (Price 1955, Gierloff-Emden 1961 and King 1972). Glaeser (1978, p. 283) noted that "only 10% of the world's barrier islands are present along coastlines where the tide ranges exceed 3 m." The question marks on Figure 1 point out areas of possible barrier island occurrence in macrotidal (T.R. greater than 4 m) areas; however, I have no detailed information on those areas.

The occurrence of barrier islands (and 6 other morphological features) relative to tidal range is illustrated in Figure 2. Note that barrier islands and river deltas are best developed in areas with tidal ranges less than 2 m; whereas, offshore linear bars (built by tidal currents), tidal flats and salt marshes are most abundant in areas with large tidal ranges (greater than 4 m). Tidal inlets and tidal deltas are more

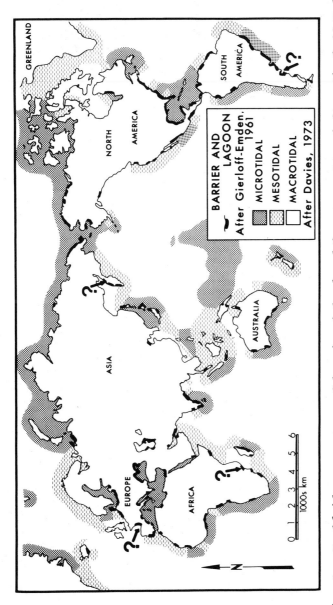

Fig. 1. Worldwide occurrence of barrier island and lagoon systems (after Gierloff–Emden 1961) in relation to tidal range (microtidal, less than 2 m; mesotidal, 2–4 m; and macrotidal, greater than 4 m; after Davies, 1973). Question marks show areas where the data from these two sources show the occurrence of barrier islands in macrotidal areas.

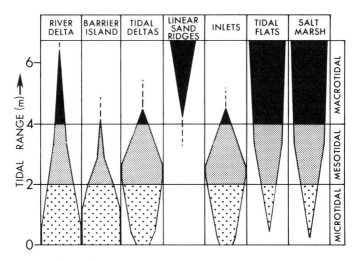

Fig. 2. Variation of morphology of coastal plain shorelines with respect to differences in tidal range (from Hayes 1975, Fig. 1).

abundant on coasts with intermediate tides (2-4 m).

Following the pioneer work of W.A. Price (1955), it is concluded that the most important control of the geomorphology of depositional coasts is the type and amount of hydrologic energy expended within an area. Furthermore, the two energy factors of most significance are wave energy and tidal current energy, which can be related directly to <u>tidal range</u>. Davies (1964) classified shorelines as follows, on the basis of tidal range:

TABLE I.

Class	Tidal range
microtidal coasts	*0-2 m*
mesotidal coasts	*2-4 m*
macrotidal coasts	*> 4 m*

In areas of average marine wave conditions, coasts with small tidal ranges (microtidal) are usually dominated by wave energy, and coasts with large tidal ranges (macrotidal) are usually dominated by tidal currents and tidal-level fluctuations. Coasts with intermediate tides (mesotidal) show influences of both waves and tides and are thus termed mixed-energy coasts (Hayes 1965).

The reason for the emphasis on tidal range is the fact that the effectiveness of wave action diminishes (i.e., waves cannot break in a concentrated area for a long period of time), and tidal current activity increases as the vertical tidal range

increases. Of course, a small tidal range does not insure high
wave energy, inasmuch as wave energy varies according to the
fetch, average velocity and duration of onshore winds in an
area. The relationship is one of mutual feedback; for example,
small waves are most effective on coasts with small tidal ranges.
On the other hand, in areas of small waves, a smaller tidal
range is required to produce tidal-influenced morphology than on
coasts with medium or high wave energy. This observation was
first made by Price (1955), who labeled the apex of the Western
Florida Bight a tide-dominated coast, even though the mean tidal
range is only on the order of 75-80 cm.

GENERAL BARRIER ISLAND SHORELINE MODELS

Introduction

   In order to simplify this discussion of morphological models,
the factor of wave energy will be held constant so that the
effect of tidal range can be examined. Only those coasts with
medium wave energy (H = 60-150 cm) [1] will be considered. It
should be pointed out that a large percentage of the barrier
island systems of the world occur in areas of medium wave energy
(refer to Fig. 15); therefore, this discussion focuses on the
more typical barrier island systems. Morphological models
(Figs. 3, 4 and 8) derived for each of Davies' (1964) three tidal classes provide a basis for discussion.

Macrotidal Shorelines

   A typical medium wave energy, coastal plain shoreline with
a macrotidal range is illustrated in Figure 3. Note the absence of barrier islands and the broad extent of intertidal
flats and salt marshes. Sand deposits are usually restricted to
linear sand shoals, or tidal-current ridges, in the offshore
areas. Examples of macrotidal coasts are found in Bristol Bay,
Alaska (Hayes and Kana, 1976); in the Wash, England (Evans 1965,
1975); at the head of the Gulf of California (Thompson 1968); at
the east entrance to the Strait of Magellan, Chile (Fischer
1977); and in many other localities (Fig. 1).

[1] *The wave height values used in this paper refer to mean
significant wave heights.*

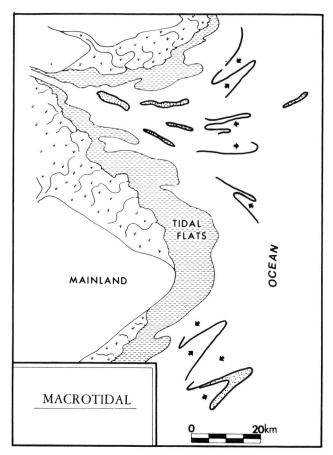

FIG. 3. Morphological model of a typical macrotidal coastal
plain shoreline with medium wave energy. Note absence of
barrier islands and presence of offshore linear sand ridges.

Microtidal Shorelines

   Introduction. Studies of microtidal barrier islands in six
geographical localities and mesotidal barriers in four localities undertaken by myself and other members of the Coastal Research Group at the Universities of Massachusetts and South
Carolina indicate that barrier island morphology is significantly different within the two tidal classes. These differences are outlined in Table 2, and the barrier island coastlines
studied are listed in Table 3.

TABLE 2. *Some general geomorphologic differences between microtidal and mesotidal barrier islands on coasts of medium wave energy (H = 60-150 cm).*

Barrier type	Length	Shape	Washover features	Tidal inlets	Flood-tidal deltas	Ebb-tidal deltas
Micro-tidal	long (30-100 km)	elong-ated hot dog	abundant; washover terraces & washover fans numerous	infre-quent	large, com-monly coupled with washovers	small to absent
Meso-tidal	stunted (3-20 km)	drum-stick	minor; beach ridges or washover terraces; washover fans rare	numer-ous	moderate size to absent	large with strong wave refrac-tion effects

TABLE 3. *Barrier island areas studied by author and associates.*

MICROTIDAL

Location	Representative publications
1. South Texas	Hayes 1965, Hayes 1967
2. Outer Banks, North Carolina	Nummedal et al. 1977, Hubbard 1977
3. West coast of Florida	Ray et al. 1973
4. Kotzebue Sound, Alaska	Ruby & Hayes, progress reports to NOAA's OCSEAP program
5. Beaufort Sea, Alaska	Ruby & Nummedal, in progress
6. SE Iceland	Nummedal et al. 1974; Ward et al. 1976; Hine & Boothroyd, 1978

MESOTIDAL

Location	Representative publications
1. South Carolina	Hayes & Kana, 1976; Hayes 1977; Finley 1978; FitzGerald 1977; Hubbard 1977; Nummedal et al. 1977
2. SE Alaska	Hayes et al. 1976; Hayes & Kana, 1976 (p. I-112 - I-120)
3. German Bight (literature only)	---
4. New England	Hayes 1969; Boothroyd & Hubbard, 1975; Hine 1975; Hubbard 1975

Morphological Model. Microtidal barriers in areas of medium wave energy are long and linear with a predominance of storm washover features (illustrated in Fig. 4). Tidal inlets and tidal deltas are of relatively minor significance. In contrast, mesotidal barriers in areas of medium wave energy are short and stunted with a characteristic drumstick shape. Tidal inlets and tidal deltas are large and significant. Therefore, probably the most fundamental difference between the two types is the presence or absence of tidal inlets. The large ebb-tidal deltas that are common on mesotidal coasts play an important role in shaping the morphology of the barriers by storing large volumes of sand which become available to the island from time to time and by strongly influencing wave-refraction patterns (discussed by Hayes 1977). Microtidal barriers are overwashed frequently by storms, probably because of a lack of conduits, in the form of tidal inlets, to allow the rising water of storm surges to flow past the barrier. Therefore, the storm surge waters break through topographic lows in the barrier islands and form washover fans. A model of washover channels and fans is given in Figure 5, and examples of washover areas from Kotzebue, Alaska and the Magdalen Islands, Gulf of St. Lawrence, are shown in Figures 6 and 7.

Mesotidal Shorelines

Figure 8 illustrates a typical mesotidal barrier island complex in an area of medium wave energy. Note the short, stunted nature of the barriers and the abundance of tidal inlets. As a rule, the ebb-tidal deltas are larger than the flood-tidal deltas (the opposite is true for microtidal areas; see Fig. 4). Some hydraulic explanations for these differences in tidal delta volumes are presented by Nummedal et al. (1977), FitzGerald (1977), and Hubbard (1977). FitzGerald suggested that the ebb dominance of the main channel at Price Inlet, South Carolina, results from greater inlet efficiency at low water than at high water, which shortens the time of ebb flow and lengthens the time of flood flow, hence producing stronger ebb currents.

Two additional aspects of mesotidal barriers tend to further distinguish them from microtidal barriers:

(1) In coastal areas with dominant waves that approach the shoreline at an oblique angle, the tidal inlets commonly show downdrift offsets; that is, the barrier beach downdrift of the inlet protrudes further seaward than the one on the updrift side (Hayes et al. 1970). The tidal inlets of New Jersey; the Delmarva Peninsula; the Copper River Delta, Alaska (Figs. 9 and 10); and South Carolina (Fig. 11) are examples.

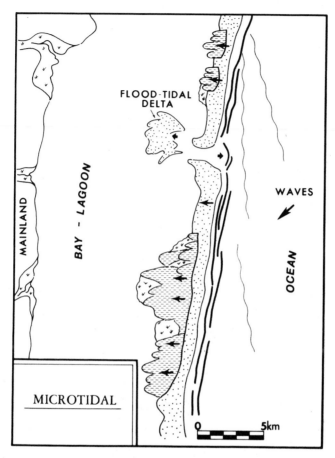

*Fig. 4. Morphological model of a typical microtidal barrier
island shoreline with medium wave energy. Note abundance of
washover areas and paucity of tidal inlets. Flood-tidal deltas
tend to be considerably larger than ebb-tidal deltas.*

(2)   Many mesotidal barrier islands have a drumstick shape,
with the fat part of the drumstick being located on the updrift
side of the barrier. Drumstick-shaped mesotidal barrier islands
from Alaska, the Netherlands, South Carolina, and Georgia are
outlined in Figure 12. Several barrier islands in South Carolina have pronounced drumstick shapes, especially Bulls Island
(Figs. 13 and 14), Kiawah Island (Hayes 1977), and Sullivan's
Island (Hayes and Kana, 1976).

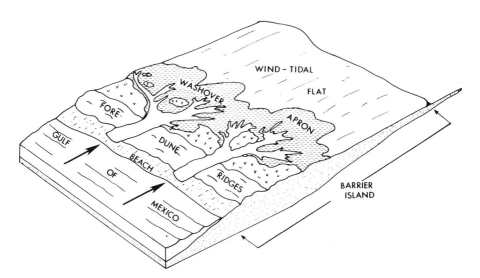

Fig. 5. Model of washover channels and washover fans for central Padre Island, Texas, a microtidal barrier in a semi-arid climate. Sediment eroded from the beach and shoreface is transported through breaches in the foredune ridge by the storm surge and deposited over the wind-tidal flats on the landward side of the island (after Scott et al. 1969).

Both the downdrift offsets at inlets and drumstick shape of mesotidal barriers are at least in part related to wave refraction around the ebb-tidal deltas, as illustrated in Figure 12A. Waves approaching the shoreline obliquely are refracted in such a way that a zone of sediment transport reversal occurs on the downdrift side of the inlet. This reversal produces a slowing down of sediment bypassing the inlet, allowing time for ridge-and-runnel systems and swash bars affiliated with the ebb-tidal delta to weld onto the beach on the downdrift side of the inlet. This process is accentuated when the main ebb channel of the ebb-tidal delta abandons a downdrift course for a more updrift one, which allows the large sand mass at the terminal lobe of the abandoned channel to be quickly driven onshore by wave action (FitzGerald 1977, Hubbard 1977). These processes combine to create the wide bulbous updrift end of the barrier, as well as the downdrift offset (Fig. 12A). The process of drift reversal at tidal inlets has been well documented in studies by Hubbard (1975) at Merrimack Inlet, Massachusetts; FitzGerald (1977) at Price Inlet, South Carolina; and Finley (1975, 1978) and Humphries (1977) at North Inlet, South Carolina.

Fig. 6.   Microtidal barrier island on west shore of Kotzebue Sound, Alaska.   Note huge washover fan in center foreground. Photograph taken in July, 1976.

Role of Wave Energy

Most of the areas studied in greatest detail by our group to date (Table 3) are areas of moderate wave energy. In order to obtain a broader sample, twenty-one barrier island shorelines, representing a wide spectrum of wave energy levels, were classified on the basis of three criteria:

(1) Mean tidal range - an average value taken from the tide tables.
(2) Mean wave height - an approximation of significant wave height based on SSMO data and the literature.
(3) Classed in one of five morphological types (purely subjective):
  a. Wave-dominated - long continuous barriers, few inlets, washovers abundant.
  b. Mixed-energy (wave dominant) - increasing numbers of tidal inlets, washovers diminishing.
  c. Mixed-energy (tide-dominant) - abundant tidal inlets, large ebb-tidal deltas, drumstick barriers.
  d. Tide-dominated (low) - occasional wave-built bars, transitional forms.

Fig. 7. Microtidal barrier spit on east shore of the Magdalen Islands, Gulf of St. Lawrence. Note abundance of washovers and absence of inlets. Photograph taken in August, 1972.

    e. Tide-dominated (high) - predominant tidal-current ridges, extensive salt marshes and tidal flats.

The twenty-one areas were then plotted on a diagram of mean tidal range versus mean wave height (Fig. 15). The values for criteria 1 and 2 (above) were determined independently by Dag

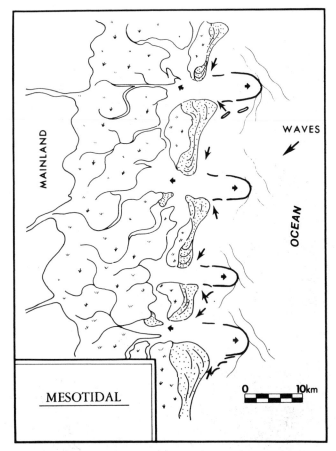

Fig. 8. Morphological model of a typical mesotidal barrier island shoreline with medium wave energy. Note abundance of tidal inlets.

Nummedal. The author made the morphological judgements (criteria 3) without prior knowledge of the detailed tide and wave data.

The results are presented in Figure 15. Note that the points converge toward the origin, with the five morphological classes radiating in bands away from the origin. The plot indicates that tidal range and wave energy effects interact to produce the characteristic morphology. For example, the central part of the Western Florida Bight (east of Appalachicola River), which has a mean tidal range of approximately 80 cm, was classified as tide-dominated (low); whereas, the southeast coast of

Fig. 9. Mesotidal barrier islands of the Copper River Delta area, Alaska. Note downdrift offsets of inlets (view looks east). Compare with map in Figure 10. Photograph taken from 5,000 ft. at low tide in May, 1975.

Iceland, which has a mean tidal range of approximately 150 cm, was classified as wave-dominated. Average wave heights at the apex of the Western Florida Bight are on the order of 10 cm; whereas, those in southeast Iceland are around 215 cm. There-fore, the obvious conclusion is that both wave energy conditions

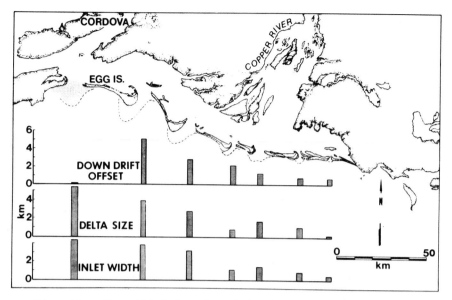

Fig. 10. Mesotidal barrier islands of the Copper River Delta area, Alaska. Four trends are apparent: (1) The downdrift offset of the inlets increases toward the west, except at the westernmost spit, which is anchored to bedrock; (2) Size of the ebb-tidal deltas increases from east to west; (3) Inlet width increases from east to west; and (4) The drumstick shape of the barriers becomes more pronounced in a westerly direction.

and tidal range must be taken into account in a meaningful explanation of barrier island morphology.

Refinement of Davies' (1964) Classification

Another conclusion of this study is that the tidal classification of Davies (1964) needs refinement. This is particularly evident if one looks in detail at the variations in barrier island morphology along the East Coast of the United States. In general, the boundary between microtidal and mesotidal (2 m) is thought to be too high. For example, New Jersey barrier islands (tidal range - 1-2 m) exhibit many obvious tidal influences. This suggests the need for a new class (here termed low-mesotidal) for areas with tidal ranges between 1-2 m. The complete modification is given in Table 4. This classification applies only to coasts of medium wave energy. More detailed data are needed for areas of low and high wave energy before attempts at refinement of the classification for those energy levels are made.

Fig. 11.   Downdrift tidal inlets on the mesotidal barrier island coast of South Carolina.   View looks southwest with Price Inlet in foreground.   Low tide photograph   by P.J. Brown.

TABLE 4.   Coastal Types - Medium Wave Energy (H = 60-150 cm)

Class	Tidal Range	Example
Microtidal	0-1 m	Gulf of St. Lawrence
Low-mesotidal	1-2 m	New Jersey
High-mesotidal	2-3.5 m	Plum Island, Mass.
Low-macrotidal	3.5-5 m	German Bight
Macrotidal	5 m	Bristol Bay, Alaska

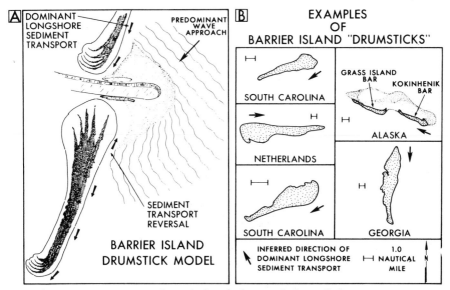

Fig. 12.   A.   *Barrier island drumstick model.*
          B.   *Examples of barrier island "drumsticks" from*
South Carolina, Georgia, The Netherlands, and Alaska.

VARIATION WITHIN SHORELINE EMBAYMENTS

   The ratio of tidal range to wave energy level tends to vary
along the coast within a shoreline embayment, with tidal range
increasing and wave energy decreasing toward the head of the embayment.   This was clearly illustrated for the Georgia Bight in
studies by Brown (1976); illustrated in Fig. 16), Hubbard (1977),
and Nummedal et al. (1977).   Accordingly, barrier island morphology changes from wave-dominated types at the entrance of the
embayment to more tide-dominated forms toward the head of the
embayment.   In fact, barrier islands commonly disappear altoge-
ther at the apex of the bay.   This mode of morphological variation is illustrated for a hypothetical embayment in Figure 17.
   Inspection of the points plotted on Figure 15 revealed that
there are different energy levels for different shoreline embayments.   For example, localities in the German Bight had higher
values of both tidal range and wave heights than those within
the Georgia Bight.   Stations within the Western Florida Bight
had even lower values.   The points within each of the three embayments roughly transcribed a tight arc across the diagram.
The three embayments are approximately plotted by the three arcs
on Figure 18.   Arc A represents the German Bight, arc B the German Bight, and arc C the Western Florida Bight.   Arc B also

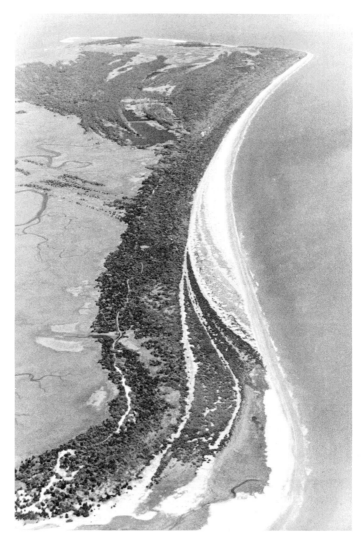

Fig. 13. Bulls Island, South Carolina, a typical mesotidal barrier island drumstick. Compare with map in Figure 14. Photograph by M.F. Stephen.

approximates the hypothetical embayment illustrated by the diagram in Figure 17. The numbers on arc B (1-4) correspond to
specific localities within the hypothetical embayment (compare
Figs. 17 and 18).

In summary, there is a continuum of shoreline morphology
around the perimeter of embayments on coastal plain shorelines
that can be related directly to interactions of wave and tidal

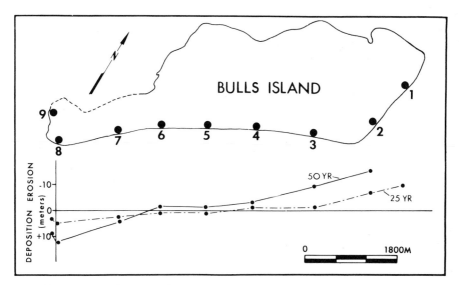

Fig. 14. Fifty and twenty-five year erosion-deposition trends at Bulls Island, South Carolina. Note that the northeastern end of the island has been strongly erosional, the central section has been fairly stable, and the southwestern end has been accretional. This pattern of erosion and deposition is typical of mesotidal, beach-ridge barriers in South Carolina (from Hubbard et al. 1977, Fig. 9).

range effects. Wave effects tend to predominate at the entrances, and tidal effects prevail at the heads of the embayments.

EFFECTS OF CLIMATE

Introduction

One more complication to these generalizations on barrier island morphology, the influence of climate, should not be ignored. In general, the ideas expressed herein apply to central latitudes (temperate climates). Not enough data are available as yet to determine precisely how much modification of the general patterns is brought about by extreme climatic conditions. A general discussion of the effect of climate on coastal morphology is presented by Davies (1973).

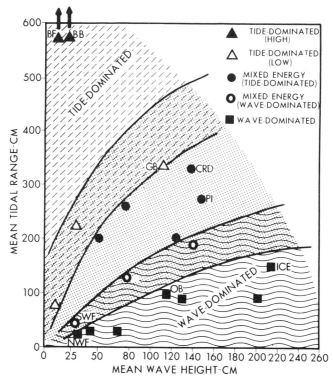

Fig. 15.   Mean wave height vs. mean tidal range for 21 coastal plain shorelines.   The areas are grouped into five morphological classes (indicated by different symbols).   BF = Bay of Fundy;   BB = Bristol Bay, Alaska;   SWF = Southwest Florida; NWF = Northwest Florida;   GB = German Bight;   CRD = Copper River Delta, Alaska;   PI = Plum Island, Mass.;   OB = Outer Banks, N.C.; and ICE = Southeast Iceland.

The Tropics

Two factors of importance in the tropics, beach rock formation and mangrove vegetation, would appear to exert some influence on barrier island morphology. Details of these effects are unknown.

A spectacular series of mesotidal barrier islands occur along the southeast shore of the Persian Gulf (Purser and Evans, 1973). These barrier islands have well-developed oolitic tidal deltas and other morphological similarities to temperate mesotidal barriers. However, cementation of beach rock and the abundance of carbonate sediments do contribute to many

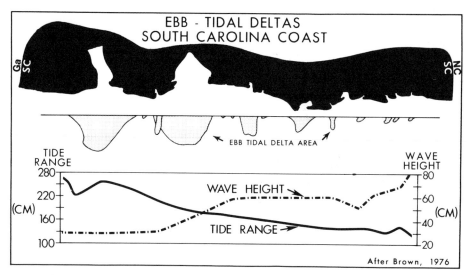

Fig. 16. Variations in wave height, tidal range and areas of ebb-tidal deltas along the South Carolina coastal segment of the Georgia Embayment. The apex of the embayment is located near the South Carolina/Georgia border. Tidal influences increase (see especially the increase in the areas of ebb-tidal deltas) and wave influences decrease from North Carolina to Georgia. Diagram by P.J. Brown.

differences in the details of morphological and sedimentation patterns between these barriers and those found in temperate regions.

Polar Regions

Some of the most extensive barrier island systems in the world occur along the microtidal shores of the Arctic Ocean. These barriers, which have been studied by Reimnitz and Barnes (1974) and Short (1976), are also presently under study by our group. Limited fetch due to ice cover, coarse gravelly sediments, and the presence of permafrost and ice gouge effects have contributed to the formation of barrier islands unlike any of the others we have studied. A typical barrier has the shape of a horseshoe. Clearly, the cold climate leaves a strong imprint on barrier morphology.

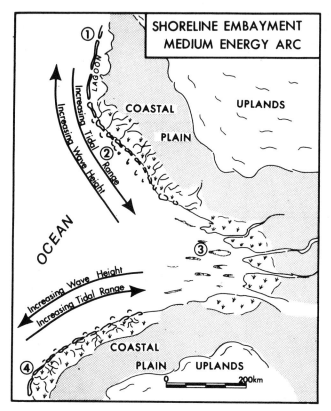

Fig. 17. Variation of shoreline morphology along an embayment in a hypothetical coastal plain shoreline with medium wave energy. The numbers represent hydrographic stations. They are plotted on the medium energy arc (arc B) in Figure 18.

Summary

Climatic effects tend to modify the impact of hydrological factors such as tidal range and wave energy on barrier island morphology. In the absence of data from polar and tropical areas, it is best at this time to assume that the conclusions of this paper apply primarily to the temperate regions of the middle latitudes.

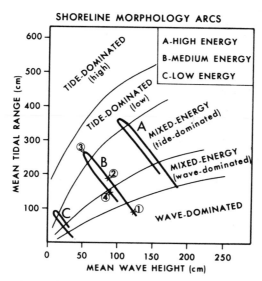

Fig. 18. *Shoreline embayment morphology arcs (A, B, and C) plotted on graph of mean tidal range vs. mean wave height. Arc A (high energy) approximates the German Bight, arc B (medium energy) approximates the Georgia Bight, and arc C (low energy) approximates the Western Florida Bight. Arc B also approximates the hypothetical shoreline embayment in Figure 17. The numbers along the arc represent hydrographic stations which are located on the map in Figure 17.*

CONCLUSIONS

(1) Barrier island shorelines are most abundant on coastal plains located on trailing edges of continents and on marginal sea coasts.

(2) Barrier islands are further restricted to those coastal plain shorelines with tidal ranges less than approximately 4 m.

(3) Field studies of 6 microtidal (T.R. < 2 m) and 3 mesotidal (T.R. = 2-4 m) areas allows the construction of general morphological models for the two tidal classes:

a. Microtidal model - Barriers long and continuous; washovers numerous; flood-tidal deltas large; ebb-tidal deltas small.

b. Mesotidal model - Barriers short and stunted with characteristic drumstick shape; washovers rare; tidal deltas large with ebb-deltas predominating.

These models relate primarily to temperate coastal areas with
medium wave heights (H = 60-150 cm).

(4)   In areas of low wave energy (mean wave ht. less than
60 cm), smaller tidal ranges are required to produce wave-
dominated coastal plain morphology than on medium wave coasts.
Conversely, larger tidal ranges are required to produce tide-
dominated morphology in areas of high wave energy (mean wave ht.
greater than 150 cm.)

(5)   For coasts with medium wave energy, a refinement is
suggested for Davies (1964) classification of tides. The boundary between microtidal and mesotidal coasts is lowered to 1 m,
and the boundary between mesotidal and macrotidal coasts is
lowered to 3.5 m.

(6)   Barrier island morphology shows distinct trends around
the perimeter of shoreline embayments. The morphology changes
from wave-dominated types at the entrance to the embayment to
more tide-dominated forms toward the head. Barrier islands
commonly disappear completely at the apex of the embayment.

(7)   Extreme climatic regimes, in polar regions and in the
tropics, appear to superimpose poorly understood morphological
imprints on barrier island morphology. Therefore, the conclusions of this paper apply only to mid-latitude, temperate coasts.

ACKNOWLEDGMENTS

Dag Nummedal is acknowledged for assistance with the wave
data and for thoughtful discussions of barrier island and tidal
inlet processes. Financial assistance was provided by a number
of sources, including: (1) Coastal Engineering Research Center,
U.S. Army Corps of Engineers (several contracts for studies in
New England and South Carolina); (2) U.S. Army Research Office
(Contract No. DAAG-29-76-G-0111, Price Inlet, S.C.); (3) Geography Programs, Office of Naval Research (several contracts for
studies in New England, Alaska and West Florida); (4) Naval Ordinance Laboratory (Cont. No. N60921-73-C-0258, Iceland); (5)
NSF-RANN Program (Cont. No. AEN-7606898, Chile); (6) NOAA's
OCSEAP Program (Cont. No. 03-5-022-82, Alaska); (7) NOAA's Sea
Grant Program, State of South Carolina (Cont. No. 04-6-148-44096);
and (8) Geology Department, University of South Carolina. The
arduous field work, painstaking data analysis, and other crea-
tive endeavors of over 40 graduate students in the Departments
of Geology at the Universities of Massachusetts and South Carolina contributed significantly to these ideas.

REFERENCES

Boothroyd, J.C., and Hubbard, D.K. (1975). Genesis of bedforms
    in mesotidal estuaries. *In* "Estuarine Research" (L.E. Cronin,
    ed.), Vol. 2, p. 217-235. Academic Press, New York.
Brown, P.J. (1976). Variations in South Carolina coastal mor-
    phology. *In* "Terrigenous Clastic Depositional Environments"
    (M.O. Hayes and T.W. Kana, ed.), Technical Report No. 11-CRD,
    Dept. Geol., U.S.C., p. II-2 - II-15.
Davies, J.L. (1964). A morphogenic approach to world shorelines.
    Zeit für Geomorph. Bd. 8, s. 27-42.
Davies, J.L. (1973). "Geographic variation in coastal develop-
    ment," 204 p. Hafner Publ. Co., N.Y.
Evans, G. (1965). Intertidal flat sediments and their environ-
    ment of deposition in the Wash. *Quart. J. Geol. Soc. 212,*
    209-245.
Evans, G. (1975). Intertidal flat deposits of the Wash, western
    margin of the North Sea. *In* "Tidal Deposits" (R.N. Ginsburg,
    ed.), p. 13-20. Springer-Verlag, N.Y.
Field, M.E., and Duane, D.B. (1976). Post-Pleistocene history
    of the United States inner continental shelf: significance
    to the origin of barrier islands. *Geol. Soc. Amer. Bull.
    87,* 691-702.
Finley, R.J. (1975). Hydrodynamics and tidal deltas of North
    Inlet, South Carolina. *In* "Estuarine Research" (L.E. Cronin,
    ed.), Vol. 2, p. 277-292. Academic Press, N.Y.
Finley, R.J. (1978). Ebb-tidal delta morphology and sediment
    supply in relation to seasonal wave energy flux, North In-
    let, South Carolina. *J. Sed. Petrology 48,* No. 1, 227-238.
Fischer, I.A. (1977). Tidal flat sedimentation in a macrotidal
    embayment, Bahia de Lomas, Strait of Magellan, Chile. M.S.
    thesis, Geol. Dept., Univ. South Carolina, 128 p.
FitzGerald, D.M. (1977). Hydraulics, morphology and sediment
    transport at Price Inlet, South Carolina. Ph.D. disserta-
    tion, Geol. Dept., Univ. South Carolina, 84 p.
Gierloff-Emden, H.G. (1961). Nehrungen und Lagunen. *Petermanns
    Geogr. Mitt. 105,* no. 2, 81-92; *105,* no. 3, 161-176.
Glaeser, J. Douglas (1978). Global distribution of barrier
    islands in terms of tectonic setting. *J. Geol. 86,* no. 3,
    283-298.
Hayes, M.O. (1965). Sedimentation on a semi-arid, wave dominated
    coast (South Texas); with emphasis on hurricane effects.
    Ph.D. diss., Univ. of Texas, 350 p.
Hayes, M.O. (1967). Hurricanes as geological agents: case stu-
    dies of hurricanes *Carla,* 1961, and *Cindy,* 1973. Inves.
    Rept. No. 61, Bur. Econ. Geol., Univ. Texas, Austin, 54 p.
Hayes, M.O. (ed.)(1969). Coastal environments: NE Massachusetts
    and New Hampshire. Cont. No. 1-CRG, Dept. Geol. Pub. Series,

Univ. Mass., 462 p.

Hayes, M.O., Goldsmith, V., and Hobbs, C.H. III (1970). Offset coastal inlets. Am. Soc. Civil Engineers, Proc. 12th Coastal Eng. Conf., 1187-1200.

Hayes, M.O. (1975). Morphology of sand accumulations in estuaries. *In* "Estuarine Research (L.E. Cronin, ed.), Vol. 2, p. 3-22. Academic Press, New York.

Hayes, M.O., and Kana, T.W. (1976). Terrigenous Clastic Depositional Environments: Some Modern Examples. AAPG Field Course Guidebook and Lecture Notes, Tech. Rept. No. 11-CRD, Coastal Res. Div., Dept. Geol., Univ. South Carolina, Part I, 131 p.

Hayes, M.O., Ruby, C.H., Stephen, M.F., and Wilson, S.J. (1976). Geomorphology of the southern coast of Alaska. Proc. 15th Coastal Eng. Conf., Honolulu, July 11-17, 1976, 1992-2008.

Hayes, M.O. (1977). Development of Kiawah Island, South Carolina. Coastal Sediments '77, Proc. 5th Symp. WPCO Div. ASCE, Charleston, S.C., Nov. 2-4, 1977, 828-847.

Hine, A.C. (1975). Bedform distribution and migration patterns on tidal deltas in the Chatham Harbor Estuary, Cape Cod, Mass. *In* "Estuarine Research" (L.E. Cronin, ed.), Vol. 2, p. 235-252. Academic Press, N.Y.

Hine, A.C., and Boothroyd, J.C. (1978). Morphology, processes and recent sedimentary history of a glacial-outwash plain shoreline, southern Iceland. Ms. submitted for publication.

Hoyt, J.H. (1967). Barrier island formation. *Geol. Soc. Amer. Bull. 78*, 1125-1136.

Hubbard, D.K. (1975). Morphology and hydrodynamics of the Merrimack River ebb-tidal delta. *In* "Estuarine Research" (L.E. Cronin, ed.), Vol. 2, p. 253-266. Academic Press, N.Y.

Hubbard, D.K. (1977). Variations in tidal inlet processes and morphology in the Georgia Embayment. Tech. Rept. No. 14-CRD, Coastal Res. Div., Geol. Dept., Univ. South Carolina, 79 p.

Hubbard, D.K., Hayes, M.O., and Brown, P.J. (1977). Beach erosion trends along South Carolina coast. Coastal Sediments '77, Proc. 5th Symp., WPCO Div. of ASCE, Charleston, S.C., Nov. 2-4, 1977, 797-814.

Humphries, S.M. (1977). Morphologic equilibrium of a natural tidal inlet. Coastal Sediments '77, Proc. 5th Symp., WPCO Div. of ASCE, Charleston, S.C., Nov. 2-4, 1977, 734-753.

Inman, D.L., and Nordstrom, C.E. (1971). On the tectonic and morphologic classification of coasts. *J. Geol. 79*, no. 1, 1-21.

King, C.A.M. (1972). "Beaches and Coasts," 2nd Ed., 570 p. Edward Arnold, London.

Nummedal, D.N., Hine, A.C., Ward, L.G., Hayes, M.O., Boothroyd, J.C., Stephen, M.F., and Hubbard, D.K. (1974). Recent migrations of the Skeidararsandur coastline, southeast Iceland. Final Report for Contract N60921-73-C-0258, Naval Ordinance Laboratory, Washington, D.C., 183 p.

Nummedal, D.N., Oertel, G.F., Hubbard, D.K., and Hine, A.C. (1977). Tidal inlet variability - Cape Hatteras to Cape Canaveral. Coastal Sediments '77, Proc. 5th Symp., WPCO Div. of ASCE, Charleston, South Carolina, Nov. 2-4, 1977, p. 543-562.

Price, W.A. (1955). Development of shorelines and coasts. Dept. Oceanography, Texas A&M, Project 63.

Purser, B.H., and Evans, G. (1973). Regional sedimentation along the Trucial Coast, SE Persian Gulf. *In* "The Persian Gulf, Holocene Carbonate Sedimentation and Diagenesis in a Shallow Epicontinental Sea" (B.H. Purser, ed.), p. 211-231. Springer-Verlag, N.Y.

Ray, P.K., Hayes, M.O., Stephen, M.F., and Ray, K.B. (1973). Multi-level cuspate features on microtidal barrier beaches. Abstract, Annual GSA Meeting, Dallas, Texas.

Reimnitz, E., and Barnes, P.W. (1974). Sea ice as a geologic agent on the Beaufort Sea shelf of Alaska. *In* "The Coast and Shelf of the Beaufort Sea" (J.C. Reed and J.E. Sater, eds.), p. 301-351. Arctic Inst. N.A., Arlington, Virginia.

Scott, A.J., Hoover, R.A., and McGowen, J.H. (1969). Effects of Hurricane *Beulah* on Texas coastal lagoons and barriers. *In* "Coastal Lagoons, A Symposium" (A.A. Castanares and F.B. Phleger, eds.), p. 221-236. Univ. Nac. Auton. de Mexico, Mex.

Short, A.D. (1976). Observations on ice deposited by waves on Alaskan Arctic beaches. *Rev. Geogr. Montr. XXX*, nos. 1-2, 1-43.

Swift, D.J.P. (1975). Barrier island genesis: evidence from the central Atlantic shelf, eastern U.S.A. *Sedimentary Geology 14*, 1-43.

Thompson, R.W. (1968). Tidal flat sediment of the Colorado River delta, northwestern Gulf of California. Geol. Soc. Amer. Mem. 107, 133 p.

Wanless, H.R. (1974). Intracoastal sedimentation. *In* "The New Concepts of Continental Margin Sedimentation,II" (D.J. Stanley and D.J.P. Swift, eds.), p. 391-429. Amer. Geol. Inst. Short Course Lecture Notes, Key Biscayne, Florida, Nov. 15-17, 1974.

Ward, L.G., Stephen, M.F., and Nummedal, D.N. (1976). Hydraulics and morphology of glacial outwash distributaries, Skeidararsandur, Iceland. *J. Sed. Petrology 46*, 770-777.

BARRIER ISLANDS
IN THE SOUTHERN GULF OF ST. LAWRENCE, CANADA

S.B. McCann

Department of Geography
McMaster University
Hamilton, Ontario
Canada

Barrier islands, barrier beaches and spits constitute over 350 km of shoreline in the Southern Gulf of St. Lawrence, a small, enclosed sea characterized by low tidal ranges, locally generated waves, and a winter period when normal shore processes are curtailed by ice. Modern studies of the barriers began in 1970, and the paper reviews the information now available about their morphology and dynamics. A central theme is the wide variety of shoreline conditions within relatively small compass. The discussion proceeds in three parts dealing in turn with the basic environmental conditions, the characteristics of the four principal barrier systems and the different sub-environments of deposition within the barriers.

INTRODUCTION

This paper reviews recent research on the morphology and dynamics of the barrier island shorelines around the Southern Gulf of St. Lawrence. There are early accounts of the barriers of New Brunswick by Ganong (1908) and in the Southern Gulf in general in D.W. Johnson's description of the New England-Acadian Shoreline (1925), but modern studies did not begin until 1970. It is perhaps not surprising then that their existence has been overlooked and that there is scant reference to them in the large barrier island literature.

The Gulf of St. Lawrence (Fig. 1) is a protected, micro-tidal, coastal environment in which normal shore processes are curtailed for more than three months each year due to the

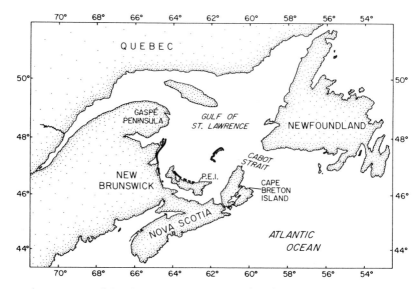

Fig. 1. Gulf of St. Lawrence, showing location of the
barrier island shorelines.

presence of pack and shore-fast ice. Swell from the North Atlantic enters the Southern Gulf for only a short time each year
and the wave climate is dominated by waves generated across the
relatively short fetches in the Gulf itself.

The barrier islands, barrier beaches and spits which have
developed in this environment across structurally controlled
embayments and estuaries are basically transgressive in character, though there are some progradational, dune ridge sequences in downdrift situations. There are four physically distinct barrier systems (Fig. 2) and a long spit, incorporating
over 350 km of shoreline, broken by some 30 tidal inlets. Table 1 summarizes the characteristics of each of the systems
and lists the principal references to each. Kouchibouguac Bay
has been studied since 1970 (Bryant 1972; Bryant and McCann,
1973; McCann and Bryant, 1970, 1973) and has provided a type
locality for the analysis of nearshore bar formation in the
Southern Gulf (Davidson-Arnott 1975; Davidson-Arnott and Greenwood, 1974, 1976; Greenwood and Davidson-Arnott, 1972, 1975).
More recent work has been concerned with a storm wave climatology for the bay (Hale 1978; Hale and Greenwood, in press)
and sediment movement across the shoreface (Greenwood and Hale,
in press).

Some initial generalizations about barriers in the Gulf as
a whole appeared in *Maritime Sediments* in 1972 (Bryant and

TABLE 1. *Barrier Systems of the Southern Gulf of St. Lawrence*

1. *Miscou Island - Point Escuminac, New Brunswick*
    a. *Miscou-Neguac (length 95 km, 12 inlets)*
        i) low beaches and barriers, modified by storm overwash, enclosing shallow lagoons; inlets small and unstable; rapid progradation in south
        ii) Owens 1974a, 1975b; Munroe 1977; Reinson, in press; Rosen, this volume
    b. *Miramichi Embayment (length 30 km, 3 inlets)*
        i) two large islands across wide estuary; large subtidal flood tidal deltas at principal inlets; extensive ebb delta shoals
        ii) Reinson 1976b, 1977b; McCann, Reinson, Armon, 1977
2. *Kouchibouguac Bay (length 29 km, 3 inlets)*
        i) simplest system with limited fetch window: temporary inlets important; nearshore bars well documented
        ii) Bryant and McCann, 1973; Davidson-Arnott and Greenwood, 1976; Greenwood and Davidson-Arnott, 1975; Hale and Greenwood, in press; McCann and Bryant, 1973
3. *North Shore, Prince Edward Island*
    a. *Malpeque (length 43 km, 4 inlets)*
        i) high dune, restricted overwash barriers; inlet related sedimentation important; sediment budget estimated
        ii) Armon, this volume; Armon and McCann, 1977, in press; McCann, Reinson, Armon, 1977
    b. *Others (length 35 km, 6 inlets)*
        i) series of small barriers and spits across small bays
4. *Magdalen Islands, Quebec*
        i. double barriers, enclosing central lagoons, connect series of small rock islands; distinct differences between east and west facing barriers
        ii. Owens 1977a, b; Owens and McCann, in press

McCann, Greenwood and Davidson-Arnott, McCann and Bryant, Owens and Harper) but two subsequent papers by Owens (1947b, 1975b) provided a fuller and more unified treatment of the different coastal environments and sediment transport patterns. Two major studies were completed in 1975 on the Magdalen Islands (Owens 1975a, 1977a, b; Owens and Frobel, 1977; Owens and McCann, in press); and on the Malpeque system in Prince Edward Island (Armon 1975; Armon and McCann, 1977a and b, in press). In 1976 attention was focused on the inlets at the entrance to the Miramichi estuary (Reinson, 1976a, b; 1977a, b) and the barriers immediately to the north (Munro 1976, 1977). Subsequently, the barriers and inlets of the Tracadie-Tabusintac section of the New Brunswick coast have been studied in some

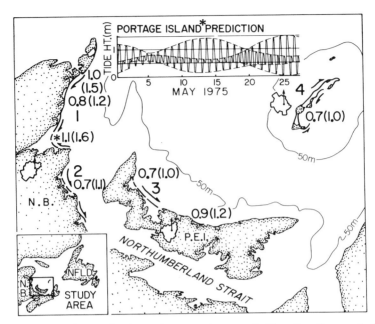

Fig. 2. The main barrier systems of the Southern Gulf of St. Lawrence: (1) Miscou Island - Point Escuminac, New Brunswick (2) Kouchibouguac Bay, New Brunswick (3) North Shore, Prince Edward Island (4) Magdalen Islands, Quebec.

The annual wind roses are for the following locations: Chatham, New Brunswick; Summerside, Prince Edward Island; Grindstone, Magdalen Islands.

The tidal information shows tidal range in meters for mean and, in brackets, large tides (Source: Canadian Hydrographic Service Tide and Current Tables). The predicted tidal curve for Portage Island (Source: Miramichi Channel Study) is typical of many stations.

Generalized longshore transport directions are shown by arrows.

detail (Huntley, in press; McCann, in press; Reinson, in press; Rosen, this volume).

In attempting to summarize and assess the information now available about the barrier coasts of the Southern Gulf of St. Lawrence, the principal impression is that there is a wide variety of shoreline conditions within a relatively small compass. One theme of this review, then, is concerned with regional variations in barrier morphology around the Southern Gulf. A second theme, open to more debate, is that the particular environmental conditions produce barrier islands which differ in some respects from the well documented barrier

shorelines further south. The differences are of degree rather
than kind, but they suggest some refinements to existing models
of barrier island sedimentation.
       The paper will proceed in three parts, dealing in turn with
the basic environmental conditions, the characteristics of the
four principal barrier systems, and the different sub-environments of deposition within the barriers--lagoon beach, dunes,
ocean beach, nearshore and offshore zones, and tidal inlets.

COASTAL ENVIRONMENT

Geological Setting

       The Gulf of St. Lawrence is an enclosed sea on the eastern
margin of North America between 46° and 50° N Latitude and 60°
and 65° W longitude (Fig. 1). It has a maximum southwest-
northeast extent of 790 km and an east-west extent of 320 km.
The barrier shorelines occur along that section of the southwest coast of the Gulf, between the Gaspé Peninsula and Cape
Breton Island, which is developed in relatively unresistant
Permo-Carboniferous sandstones. Coastal relief is low and the
broad estuaries and embayments are considered to represent a
drowned preglacial drainage system which continues seawards
across the Shallow Magdalen Shelf (Kranck 1972a; Loring 1973;
Owens 1974b). Pleistocene ice sheets did not greatly modify
this drainage pattern and the principal effect of glaciation
was the deposition of predominantly gravelly and sandy tills
and moraines across the shelf and adjacent land area. The details of the Pleistocene history of the region are still subject to controversy (see Prest et al. 1976, which provides many
additional references) and are not particularly relevant to
this discussion. It is important to note that there is an abundant source of sandy sediments across the Magdalen Shelf, which
have been reworked during the postglacial marine transgression.
Loring and Nota (1969, 1973) indicate that the surficial sediments of large areas of the Shelf are undergoing active rework-
ing and redistribution at the present time. Sands are being
carried shorewards and fines are being transported to deeper
water. Recent sea-level history throughout the region is one
of sea level rise of the order of 3 m/ 1,000 y (Farquharson
1959; Frankel and Crowl, 1961; Grant 1970; Kranck 1972; Thomas
et al. 1973) but there are indications that there have been,
and continue to be, variations from place to place. Kranck
(1972) suggests that there are differences between the eastern
and western sections of the Northumberland Strait in terms of
postglacial sea level changes, and Vanicek (1976) suggests that
there is some evidence of recent uplift in northeastern New
Brunswick.

Climate

The important factors in the climate of southern Gulf, in
terms of coastal processes, are the cold winters and the wind
regime. Mean monthly temperatures are below 0^{0}C for the period
December to March (Fig. 3A), and sea ice, which begins to form
in late December, commonly remains until late April (Forward
1954; Owens 1974b). Fast ice protects the beach from wave action, and pack ice offshore inhibits wave generation for up to
three months each year, during the stormiest period of the year.
Figure 3B illustrates the duration of ice protection on the
north coast of Prince Edward Island during the period 1959-70.
The inner parts of the lagoons are usually frozen over by late
December, and the beach is totally protected from wave action,
on average, between January 15 and April 25. Sea surface temperatures (Fig. 3C) are below 0^{0}C from mid-December to late
March. Ice thicknesses in the lagoons commonly reach 0.5 m,
and some of the smaller lagoons may be frozen to bottom, with
the reduced tidal prisms affecting inlet processes. Most wri-
ters have stressed the inhibitive role of ice, and regard the
various beach features caused by ice push, ice rafting and ice
melting as ephemeral ones, which are rapidly removed when more
normal wave processes commence in the spring. There are no
comparative data on the effects of ice in the different barrier
systems and little consideration has been given to the role of
ice in overall barrier development. The fact that wave processes do not occur across the beach zone in winter clearly reduces total annual wave energy levels at shore in comparison with
ice-free barriers further south, but more important is the reduced frequency of severe winter storm wave action (Armon and
McCann, in press).

A further climatic effect, which has received little attention, is the relatively late start of the growing season. Sand
dune vegetation, in particular the colonizers of the mobile
dune, does not show any appreciable new growth until mid-June.
The late spring-early summer period is relatively dry and aeolian sand transport is important at this time of year.

The weather patterns and wind regime of the Southern Gulf
are related to the passage of mid-latitude low pressure cells
(depressions) across the area, and the prevalent winds are westerly. The annual wind roses for Chatham and Summerside (Fig.2)
indicate the prevalence of southwesternly winds on the western
section of the Gulf. These are offshore winds for the New
Brunswick and Prince Edward Island barriers, but there is a secondary peak from the northeast, which is the direction of maximum fetch. The annual wind rose for Grindstone, in the Magdalen

Islands in the centre of the Gulf, shows an increase in winds from the northwest. There is a marked seasonal variability in wind speed and direction throughout the year. Figure 4 shows the 9-month wind rose (i.e., ice-free period, April - December) for Summerside, and four "seasonal" wind roses. November and December are the stormiest months during the ice-free year, with a dominance of winds from the west, but April and May show a more pronounced peak in moderate and strong winds from the northeast. Owens (1975a, 1977a) selected August and November as representative summer and winter months, respectively, for his seasonal observations of meteorological and wave conditions in the Magdalen Islands in 1974. He reports a very consider-able increase in wind speeds, and correspondingly, in wave heights and wave energy levels in November.

Waves

There has been no comprehensive study of the wave climate of the Southern Gulf, but a compilation of data from different sources allows some generalizations to be made and some compari-sons to be drawn between different barrier situations. Near-shore wave observations associated with shoreline studies have been made at Kouchibouguac Bay (Greenwood and Davidson-Arnott, 1975) and in the Magdalen Islands (Owens 1975a, 1977a). Syn-optic, local, deep water wave climates have been derived for Malpeque (Armon 1975; Armon and McCann, 1977a) and Kouchibou-guac Bay (Hale 1978; Hale and Greenwood, in press). The data for Malpeque provided a basis for wave refraction analysis to determine inshore wave conditions and longshore transport rates (Armon and McCann, 1977b). Real data on deep water wave condi-tions is limited: it is derived from two sources--a summary of shipboard observations (Atmospheric Environment Service, 1972) and accelerometer buoy records for various locations (Ploeg 1971; Ashe and Ploeg, 1971; Marine Environmental Data Service, ongoing reports). There is also a hindcast compilation of characteristic hourly wave lengths and periods for the central Gulf for the months March to December, 1956-1960 (Quon et al. 1963).

Three things are evident from this heterogeneous collection of information. Firstly, the deep water wave climate of the Southern Gulf is dominated by short period waves, generated within the Gulf itself. Secondly, the seasonal variations in wind conditions result in a similar variation in wave condi-tions, though these are masked to some extent by shorter term variations, related to the passage of individual depressions. Thirdly, there are considerable differences in wave energy levels between different barrier systems, which result from shoreline orientation in relation to prevalent wind directions,

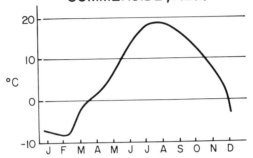

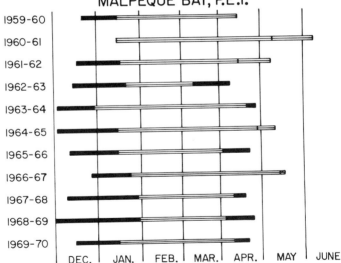

Fig. 3. (A) Mean monthly temperatures at Summerside, P.E.I. (B) Period of beach protection by ice at Malpeque, P.E.I. 1969-70. Solid shading indicates lagoon frozen at Bideford; open shading indicates fast ice present on beaches or pack ice present offshore. (C) Sea surface temperatures, S.W. Gulf of St. Lawrence, 1962-1971 (Source: Armon 1975).

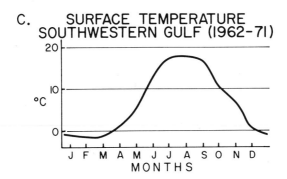

Fig. 3C

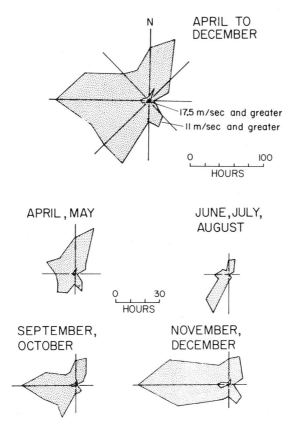

Fig. 4. Nine-month and "seasonal" wind roses for Summer-side, P.E.I. (Source: Armon and McCann, 1977a).

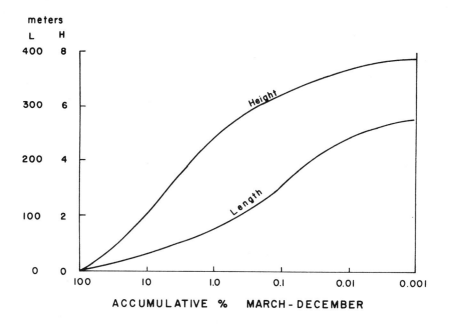

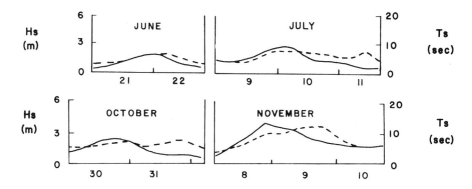

Fig. 5 (A) Wave conditions in the central Gulf of St. Lawrence (after Quon et al. 1963). (B) Wave characteristics for four selected storms recorded at North Point, P.E.I. (after Armon and McCann, 1977; data from Ploeg 1971).

as well as more local wave refraction effects. Atlantic Ocean
swell enters the Gulf for only about 7% of the year, and is
important only on the east coast of the Magdalen Islands. The
formation of true swell waves over the short fetches in the
Gulf is unusual, and waves are steep with short wavelengths in
relation to wave heights. Wave periods greater than 9.5 s and
wave heights greater than 1 m are uncommon (Fig. 5A), and the
larger waves associated with individual storms are short-lived
(Fig. 5B). The importance of onshore winds from very particular fetch directions is illustrated by the summer nearshore
wave and wind data for the Magdalen Islands and Kouchibouguac
Bay (Fig. 6).

Kouchibouguac Bay, with a very restricted fetch window to
the northeast (McCann and Bryant, 1973), experiences the lowest
wave energy conditions of the four main barrier systems. In
the context of this bay, Hale (1971) defines a storm event as
a period during which winds, from between north and southeast,
blow at speeds greater than 19 km/h for at least six hours.
His analysis indicated that there are, on average, 19 such
events each year and that waves greater than 1 m high occur
once every 44 days on the average. May is the stormiest month.
In contrast, the west coast of the Magdalen Islands is probably
the highest energy barrier shore with November and December being the stormiest months (Owens 1977b). The selected nearshore
wave information shown in Table 2 encompasses the range of conditions likely to be encountered on the barrier beaches of the
Southern Gulf.

Tides and Tidal Currents

The Gulf of St. Lawrence is too small for any large response
to tide generating forces, and the tide propagated through Cabot
Strait and Belle Isle Strait from the North Atlantic is of a
mixed type, though mainly semi-diurnal, except along certain
parts of the Northumberland Strait and near Savage Harbour,
P.E.I., where diurnal inequalities dominate (Farquharson 1968;
Dohler 1969. In general the Gulf is a microtidal environment,
with mean tidal range less than 2 m; and most of the barrier
island shorelines experience tidal ranges during spring tides
of less than 1.6 m (Fig. 2). The predicted tidal curve for
Portage Island at the entrance to the Miramichi estuary, also
shown on Figure 2, illustrates the mixed semi-diurnal and diurnal nature of the tides. At the period of neap tides, two
nearly equal high and low waters occur per day with a range
less than 0.5 m; but at spring tides there is only one prominent high and low water per day, with a range of 1.5 m, and a
small secondary oscillation on the tidal curve. Tidal currents,
away from tidal inlets, are generally weak, except in the

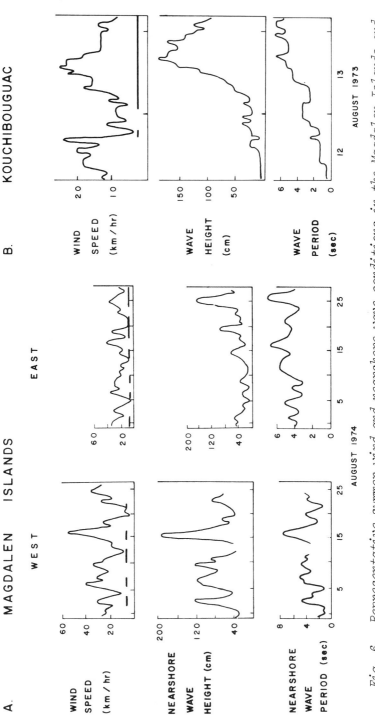

Fig. 6. *Representative summer wind and nearshore wave conditions in the Magdalen Islands and at Kouchibouguac Bay (A. after Owens 1977b; B. after Greenwood and Davidson-Arnott, 1975).*

40

TABLE 2. Wave Conditions at Selected Localities in the Gulf of St. Lawrence

NEARSHORE	cm.	cm.	sec.	sec.
	Average Wave Height	Maximum Wave Height	Average Wave Period	Maximum Wave Period
Kouchibouguac Bay[a]				
Summer	36	189	2.8	6.4
	Average Significant Wave Height	Maximum Significant Wave Height		
Magdalen Islands, East[b]				
Summer	35	200	4.1	
Winter	98	350	5.9	7.5
Magdalen Islands, West[b]				
Summer	49	230	4.1	
Winter	110	392	5.1	8.0
OPEN GULF	Median Significant Wave Height	1% Significant Wave Height	Average Wave Period	1% Wave Period
Bird Rocks[c]				
Summer	107	335	5.4	9.4
Winter	189	564	7.0	12.6

Sources:
[a]Greenwood and Davidson-Arnott, 1975 (Aug. 9 - 23, 1973).
[b]Owens, 1977a (July 29 - Aug. 25, 1974 and Nov. 14 - 30, 1975).
[c]Ploeg, 1971 (July 27 - Sept. 15, 1967 and Nov. 1 - Dec. 11, 1967).
Bird Rocks are located 25 km N of the Magdalen Islands, with long fetches to the NW and NE, and open to the influence of Atlantic Swell.

Northumberland Strait where the distribution of bottom sediments is controlled mainly by tidal currents (Kranck 1972b).

BARRIER SYSTEMS

Miscou Island to Point Escuminac

     The barrier islands, barrier beaches and spits along the
northeast Coast of New Brunswick (Fig. 7A) may be divided on
morphological grounds into two units: the Miscou-Neguac unit in
the north and the Miramichi unit in the south. The northern
unit is characterized by low beaches and barriers, less than
3 m above MHW, which are frequently modified by storm overwash
processes (Rosen, this volume). The beaches enclose a series
of small, shallow lagoons and are breached at the present time
by 10 inlets. All the inlets are small and most are unstable,
with a tendency to migrate and/or close periodically (Reinson,
in press). Littoral drift is directed southwards, with only
local reversals; and at the southern end of Neguac (Fig. 7B),
very rapid inlet migration and spit extension has occurred in
recent years (Owens 1975b; Munro 1977).
     Neguac Island marks the northern entrance to the broad,
funnel-shaped Miramichi estuary, which is 22 km wide at the
mouth and extends 80 km inland to the tidal limit. On the south
shore of the embayment, west of Point Escuminac, littoral drift
is directed to the west and north (Owens 1974a, 1975b). Two
large barrier islands, Portage Island and Fox Island, have developed across the entrance of the embayment, which acts as a
large sediment sink for the southward littoral drift through
the Neguac section and for material westward along the southern
shore (Owens 1975b; Reinson 1976b). The Miramichi river, with
a drainage basin of 14,000 km^2, also contributes a considerable
volume of sediment to the estuarine system. Portage and Fox
Islands are much larger and more stable than the barriers further north, and include a series of well-vegetated dune ridges.
Fox Island has been particularly stable over the period of historical record since 1837, the most important change being the
closure of a small inlet, Huckleberry Gully, at the southern
end in 1972. Portage Island changed considerably in the period
1837-1922. losing some 3 km at the northern end and extending
southeastward (Willis 1977). These and subsequent changes are
related to changes in the two principal inlets, Portage Gully
and Portage Channel (Reinson 1976a, 1977b), which will be discussed in the section below dealing with inlets. In the present context, however, it is interesting to speculate on the
evolution of Portage Island prior to the first historical record
in 1837. It appears that the original terminal point of a long

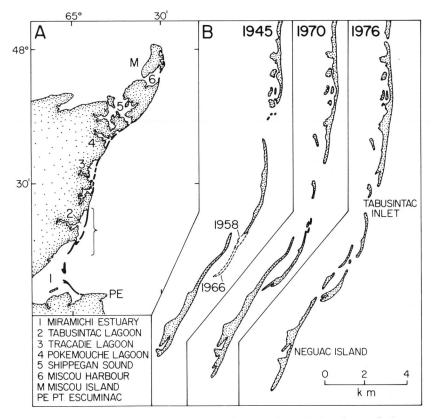

Fig. 7 (A) The New Brunswick Barrier islands and lagoons between Miscou Island and Point Escuminac. (B) Shoreline changes between 1945 and 1976 in the Tabusintac-Neguac segment (in part after Owens, 1974, and Munro, 1977).

spit was an extension of barriers further north. This suggestion is supported by the orientation of the sequence of old (pre-1837) dune ridges in the island, and by analogy with the present condition of Buctouche spit to the south. The proximal section of this long recurved spit, which is very low and narrow, is a zone of sediment bypassing: the distal section consists of multiple, recurved ridges and has an overall shape similar to Portage Island in 1837.

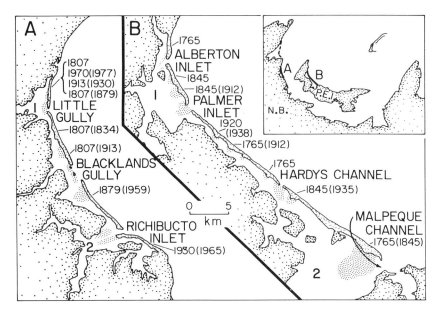

Fig. 8 (A) Kouchibouguac barrier system, New Brunswick.
(B) Malpeque barrier system, P.E.I. Both maps show the posi-
tion of present and former inlets. Dates of inlet formation
and closure (in brackets) from Bryant and McCann (1972), and
Armon and McCann (in press).

Kouchibouguac Bay

   This system consists of 29 km of sand beaches and dunes in
a gentle arc across the bay, broken by three inlets of which
the southerly one, Richibucto Inlet, is the largest (Fig. 8A).
It occupies the most protected barrier situation in the Southern Gulf, with only a very narrow fetch window to the northeast.
Littoral drift is directed southward and the islands increase
in width, height and complexity in this direction. The transgressive nature of the system is indicated by the presence of
lagoonal peat deposits near low tide level on the seaward foreshore, though not all sections of the shoreline have undergone
significant retreat in the period since 1930 (Bryant and McCann,
1973). The three permanent inlets (since 1807) are located near
the widest parts of the lagoons, opposite the drowned estuaries.
Several temporary inlets have occurred at other locations (Fig.
8A) and the most recent, opened in the winter of 1970-71 at the
northern end of the system, has already closed. At least three
previous inlets have breached the northern spit since 1807 and
this area is frequently overwashed in storms. The temporary inlets further south have generally been larger than the northern

ones and their locations are marked today by relict, flood tidal deltas. The seaward dune ridge is also a more substantial
feature in the southern two-thirds of the system, reaching 6-7
m above MHW, and there is a small progradational sequence north
of Richibucto Inlet. In many respects the system is a less dynamic and more subdued analogue of the Malpeque System, described below.

Malpeque Barrier System

Barrier islands, spits and beaches occur across most of the
estuarine embayments along the north coast of Prince Edward
Island (Fig. 8B), but only the 43 km long Malpeque system in
the west has been studied in detail (Armon 1975). Net longshore transport is directed to the southeast at Malpeque, at
rates varying between 40,000 and 2000,000 m^3y^{-1}, and subtidal
erosion of the shoreface contributes much of the available sand
(Armon and McCann, 1977). The whole system exhibits an erosional tendency at the present time, with an average rate of
dune line retreat of 0.26 my^{-1} in the period 1935-1968 (Armon
and McCann, in press). However, progradation has occurred in
the south at some time in the past 1,500-2,000 years to develop
the dune ridge sequence on Hog Island (McCann 1972), and landward sediment transfers across the barrier at the present time
are maintaining the equilibrium form of the system during retreat (Armon and McCann, in press). Most of these landward
transfers of sediment are taking place at the sites of present
and former inlets, and Armon (this volume) provides a fuller
discussion of the significance of these transfer processes.
There are presently three distinct islands and a spit at Malpeque, but on occasions during the past 200 years there have
been as many as six islands and seven inlets (Fig. 8B). Malpeque Channel in the south is the largest inlet, comparable in
scale and associated sand bodies with the Miramichi Inlets.
Hardy's Channel (Armon 1979, in press) is typical of many of
the smaller inlets in the Southern Gulf. An important characteristic of the Malpeque system is the high and relatively continuous seaward dune, often in locations where no other dunes
are present, which restricts overwash to the sites of former inlets. The lobate lagoonal margin of much of the system is related to sediment interception at these inlet locations, by tidal currents when they were open and by overwash following
their closure.

Magdalen Islands

The Magdalen Islands occupy a unique position on the summit of the shallow, sediment-abundant, Magdalen Shelf in the centre of the Gulf of St. Lawrence (Fig. 9). Two large tombolo systems of double barrier beaches and dunes, enclosing extensive, shallow lagoons, join a series of small bedrock islands, and together with complex, terminal spits, make up a continuous, narrow land area 70 km long (Fig. 1). The islands may be subdivided into nine geomorphic units, five of which are bedrock units with coastal cliffs, and four of which are depositional-barriers and spits (Owens and McCann, in press). Two of the bedrock units remain as isolated islands, but the other three act as stable rock anchors for the tombolo systems and the terminal spits. The barrier beaches and dunes are composed of relatively uniform medium-sized sands, derived largely from re-working of the sandy glacial sediments of the Magdalen Shelf by beach and shallow marine processes over a long period of fluc-tuating, but overall rising sea level. These processes con-tinue today. The Magdalen Islands provide the only barrier lo-cation in the Gulf where the prevalent westerly winds cross relatively long fetches. This means that locally-generated waves out of the west and northwest determine the longshore sediment transport patterns along the islands (Owens 1975b). This leads to differences both in the dynamics and morphology of the west and east-facing beaches (Owens 1977a; Owens and Frobel, 1977) and in the overall character of western and eas-tern barriers (Owens and McCann, in press). In general the west-facing beaches are zones of sediment bypassing at the pre-sent time: this is particularly the case along the northern tombolo, where Owens (1977a) calculated a net littoral drift to the north at the rate of 234,000 m^3y^{-1}. The western barriers of both tombolos are narrow, and in the south there is landward migration into the central lagoon. In contrast, sections of the eastern barriers contain wide, progradational, dune ridge sequences.

Discussion

As the preceding paragraphs have shown, there is consider-able variation in the morphology and structure of the barrier islands and spits around the Southern Gulf of St. Lawrence. However, comparison of the different barrier systems allows some broad generalizations to be made. Most of the barriers are undergoing slow shoreline retreat at the present time but appear to be maintaining an equilibrium form in the process. Three of the systems (Miscou Island-Point Escuminac, Kouchibou-guac and Malpeque) occur on west-facing or "lee" shores, and

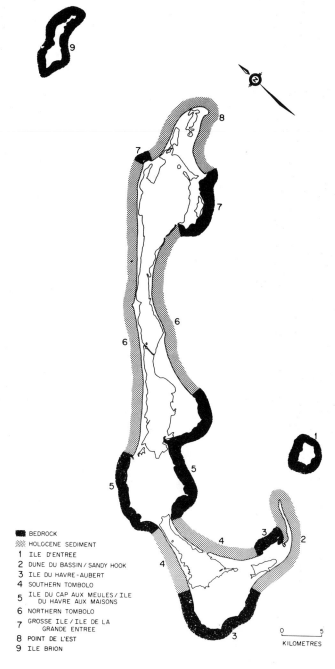

BEDROCK
HOLOCENE SEDIMENT
1 ILE D'ENTREE
2 DUNE DU BASSIN / SANDY HOOK
3 ILE DU HAVRE-AUBERT
4 SOUTHERN TOMBOLO
5 ILE DU CAP AUX MEULES / ILE
   DU HAVRE AUX MAISONS
6 NORTHERN TOMBOLO
7 GROSSE ILE / ILE DE LA
   GRANDE ENTREE
8 POINT DE L'EST
9 ILE BRION

0   5
KILOMETRES

Fig. 9.   Geomorphological subdivisions of the Magdalen
Islands, Quebec (after Owens and McCann, in press).

net longshore transport is directed to the south under the influence of winds and waves from the northeast. There is accordingly an overall increase in the dimensions of the barrier
sand bodies from north to south in the three systems. In each
case the largest lagoon and estuary and principal inlets occur
at the southern end, where the associated flood tidal deltas
and ebb-tidal delta-shoal complexes attain their greatest development. In addition, progradational dune ridge sequences
have formed sometime in the past at the southern end of the
systems--on Portage and Fox Islands at the entrance to the Miramichi estuary, north of Richibucto Inlet at Kouchibouguac, and
on Hog Island at Malpeque. Because of the drift reversal, west
of Point Escuminac, the Miramichi embayment and estuary acts as
a sediment sink for sand moving southwards in the Miscou-
Escuminac system, but there is an export of sand at the southern
ends of the Malpeque and Kouchibouguac systems. In the Magdalen
Islands, in the centre of the Gulf, waves from the west and
northwest are the most important in determining the general
pattern of sediment transport, and sand is being moved from the
west coast towards the terminal spits. The largest progradational dune ridge sequences in the southern Gulf occur along
the sheltered east coast of the islands.

DEPOSITIONAL ENVIRONMENTS

Most research to date has been concerned with the morphology
and processes of the ocean beach, nearshore zones and tidal inlets. Little attention has been focused on aeolian processes
or the role of vegetation in either dunes or salt marshes.
Though the deeper shelf sediments in the Southern Gulf are well
documented, little is known regarding patterns of sediment
transport in the inner shelf-shoreface zone.

Lagoon Beach

Apart from the major drowned estuaries, the lagoons at Kouchibouguac, Malpeque, and along the Miscou-Neguac section of
the New Brunswick shore are narrow and shallow. Widths less
than 2 km are common and depths less than 1 m at high tide prevail over large areas distant from tidal distributary channels.
Thus, many kilometers of lagoonal shoreline are very low energy
environments, but they display a wide range of depositional conditions. Straight sections of sand beach, often backed by a low
foredune, are broken by relict and active washover fans, relict
flood tidal deltas and stretches of narrow, fringing salt marsh.
Grandtner (1966) identified four major vegetation communities

between the enclosed lagoons and the dunes in the Magdalen
Islands. On the margin of the lagoons, *Salicornia europaea*
and *S. laurentium* give way to *Spartina alterniflora* in the upper part of the intertidal zone, followed by a landward progression from *Carex palacea* to *Juncus balticus*. Other salt
marsh species at Kouchibouguac include *Limonium carolinianum*,
Solidago sempervirens and *Plantago oliganthus* (McCann et al.
1973). *Spartina patens* occupies high marsh situations in
Prince Edward Island.

Dunes

There is considerable variation in the character of the
supratidal zone of the barriers of the Southern Gulf, which is
well illustrated by comparing the high dune shoreline at Malpeque with the low overwashed condition of the Miscou-Neguac
barriers. Along much of the Malpeque shore the dune zone is
only 200-400 m wide, but it is dominated by a high seaward dune,
8-10 m high above LLW. This seaward dune which may be twice
the height of the dunes further inland and also occurs in situations where no other dunes are present, appears to be maintaining its dimensions during overall barrier retreat (Armon
and McCann, in press). Similar high dune shore conditions occur
in the southern part of the Kouchibouguac system (Bryant and
McCann, 1973) and along the western barrier of the northern
tombolo in the Magdalen Islands (Owens and McCann, in press).
Along the Miscou-Neguac shore, as exemplified by the Tabusintac
section (Rosen, this volume), the supratidal zone is only 120 m
wide, average elevations are only 3 m above LLW, and dune ridges
are not continuous. Storm overwash and wind transport in overwash passages are the important processes here.
The reasons for these differences in dune character and
elevation are not yet clear. Owens (1974a, 1975b) suggested
that there is a relationship between barrier elevations and the
nature of the associated longshore sediment transport system.
He considered that low barriers with poorly developed dunes are
associated with relatively high rates of longshore transport,
and that well developed dunes occur where longshore transport
rates are low. This generalization cannot be substantiated in
the light of subsequent determinations of longshore transport
rates for Malpeque (Armon and McCann, 1977) and the west coast
of the Magdalen Islands (Owens 1977a). Both are high dune situations with high rates of longshore transport, that may well exceed the rates along the Miscou-Tabusintac shore where no reliable estimations have been made. Other important factors to be
considered are availability of sediment, shoreline orientation

in relation to wind regime, the frequency and magnitude of storm surges, and the rate of barrier retreat in the different barrier situations.

Progradational dune ridge sequences occur at several locations, being best developed in the east coast barriers of the Magdalen Islands. Recurved beach ridges with low dunes occur at the terminus of the complex, prograding Buctouche spit and in the terminal spits of the Magdalen Islands.

Sand dune vegetation has been studied in the Magdalen Islands, at Brackley in Prince Edward Island, and at Kouchibouguac. In the Magdalen Islands (Grandtner 1966, 1968; Lamoureux and Grandtner, 1977) the mobile dunes, as elsewhere, are dominated by *Ammophila breviligulata*, but this is replaced in the fixed dunes of the progradational ridge sequences by *Emperrum nigrum*. Zones dominated by *Cladonia*, *Juncus* and *sphagnum* are also present within the old dunes, and *Hudsonia tomentosa*, *Juniperus communis* and *Myrica pensylvanica* are also common. Grandtner (1970) defined a progression of plant communities at Brackley as follows:

(1) The mobile dunes, characterized by *Ammophila*, *Cakile edentula*, *Solidago sempervirens*, *Oenthera sp.*, *Carex silicea*.

(2) The fixed dunes, in which shrubs, such as *Myrica* and *Hudsonia*, and various lichens and mosses are replaced by white spruce *(Picea glauca)*-bayberry forest. As the spruce forest closes, the bayberry disappears and there is an increase in forest herbs, mosses and lichens.

(3) The wet depressions (frequently old blowouts), in which *Juncus balticus* precedes invasion by *Myrica* and *Spirea latifolia*, and subsequently other shrubs, willows, grey birch and hygrophilous herbaceous species.

The situation at Kouchibouguac (Mc Cann et al. 1973) is probably more typical of most of the Southern Gulf than either the wide progradational dune sequences of the Magdalen Islands or the old high dunes at Brackley. There are only very small areas of fresh water marsh and only one isolated blowout colonized by *Myrica gale*, *Rosa nitida*, *R. carolina* and *Fragaria virginiana* surrounded by a fringe of *Arctostaphylos uva-ursi*. For the most part, the dune vegetation of the barrier islands and spits at Kouchibouguac consists of *Ammophila*, *Lathyrus japonicus*, *Hudsonia*, and the lichen *Cladonia alpestris* and *C. rangferina* present to lesser degrees. *Ammophila* together with *Cakile edentula* predominates on the more mobile dunes and foredunes. *Hudsonia*, together with lichens, is important in some sheltered areas in the centre of the islands.

Ocean Beach

With the low tidal ranges in the Southern Gulf the intertidal beach zone is usually narrow (40-50 m at Kouchibouguac;
30-70 m at Malpeque; 25 m and 45 m, respectively, on the western and eastern sides of the Magdalen Islands). There are,
however, considerable variations alongshore, and shoreline un-
dulations or protuberances (Owens 1977a), also variously described as giant cusps (Komar 1971; Greenwood and Davidson-
Arnott, 1975) or beach rhythms (Dolan 1971; Armon 1975), occur
on most beaches. Cusp wavelengths tend to be different in different environmental settings (90-300 m at Kouchibouguac,
75-400 m at Malpeque, 200-250 m on the east side of the Magdalen
Islands) but never reach the size of the features described by
Dolan (1971) on the Cape Hatteras coast, where cusp wavelengths
of 500-600 m are common. The size of the ridge and runnel systems, which characterize the inner nearshore zone of many of
the beaches, are also smaller than those described for more open
ocean coasts. Ridges are commonly 25 m wide and 20 cm high at
Malpeque, 25-30 m wide and 25-30 cm high on the east coast of
the Magdalen Islands, and 40-50 m wide and 75 cm high on the
west coast of the Magdalen Islands (cf. ridges 70 m wide and
1 m high described by Hayes and Boothroyd, 1972, from Massachusetts). However, the landward migration of the low ridges across
the low tide terrace during post-storm beach recovery tends to
be more rapid than in meso-tidal environments (Owens and Frobel,
1977). The sequence of events in beach recovery, following
storm wave erosion in the Magdalen Islands, documented by Owens
and Frobel, involves initial modification of the profile by the
rapid growth, landward migration and welding of a small lens of
sediment during the first two or three days after a storm, and
the subsequent formation of a larger ridge on the low tide terrace, which may take a week or more to weld onto the beach.

Nearshore Zone

The basic environmental conditions for the development of
nearshore bars--gentle offshore slope, small tidal range and
limited fetches--occur in all the barrier systems of the Southern Gulf. Consequently, bar topography is a ubiquitous and
permanent feature of the nearshore zone. Number and morphology
of the bar systems depends on wave energy conditions and sediment availability. There are differences in bar characteristics within, as well as between, the barrier systems, but any
given section of beach appears to have a characteristic pattern
which is maintained from year to year. Armon (1975), for instance, reports variations in the number of bars and the widths
of the bar zone along the Malpeque system. The nearshore zone,

A. KOUCHIBOUGUAC

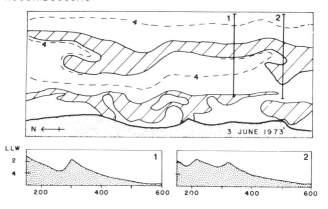

B. MAGDALEN ISLANDS (WEST)

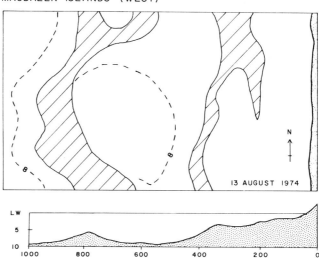

Fig. 10. Maps and profiles of nearshore bar systems in the southern Gulf of St. Lawrence. (A) Kouchibouguac Bay (after Greenwood and Davidson-Arnott, 1975). (B) Magdalen Islands, west coast (after Owens 1977). (C) Magdalen Islands, east coast (after Owens 1977).

On the maps submarine contours are in meters, bars are shown by diagonal shading, and the beach by stippling.

C.   MAGDALEN ISLANDS (EAST)

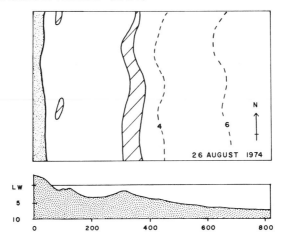

Fig. 10C

incorporating two or three bars, extends between 400-700m seawards of the waterline, and the junction between the seaward
slope of the outer bar and the low angle shoreface occurs in
depths between 3-5 m. The system of bars is widest, 580 m mean,
and the depth of the outer bar crest greatest, 3.6 m mean, in
the south where wave energy levels are highest. Further north
the mean width of the bar zone is only 380 m and the mean depth
of the outer bar crest only 1.5 m.

   The morphology and sedimentology of the nearshore bars is
very well documented in the relatively sheltered environment
of Kouchibouguac Bay (Davidson-Arnott 1975; Davidson-Arnott and
Greenwood, 1974, 1976; Greenwood and Davidson-Arnott, 1972,
1975), where there are usually two distinct bar systems. There
is usually only one outer bar, which forms a continuous pattern
of crescents, some 200-300 m offshore, in minimum water depths
of 1.8 m at low tide (Fig. 10A). The inner bar system (extending up to 130 m seawards of the low tide water line) is more
complex and some areas are emergent at low tide. All waves
greater than 0.5 m high break on these bars at all stages of
the tide, but the outer bars are only affected by a limited
spectrum of waves generated by strong northeasterly winds.
Greenwood and Davidson-Arnott (1975) maintain that both bar systems are formed by the same basic set of processes associated
with breaking waves. Seaward transfer of sediment in the inner
bar system is accomplished by rip currents and a distinctive

rip channel facies occurs in the bar sediments. Seaward transfer of sediment in the outer bar system is also related to uni-
directional currents which flow seawards across the bar crescents, but are not confined to distinct channels. Direct observation of these currents, which occur during storms, is diffi-
cult, but evidence of their existence is provided by seaward-
dipping megaripple units within the bar crest facies (Davidson-
Arnott and Greenwood, 1976).

In the more dynamic environment of the west coast of the
Magdalen Islands (Owens 1977) there are three bar systems--a
crescentic outer bar, a straighter middle bar and a discontinuous inner bar (Fig. 10B). The outer bar crest is between
700-1000m offshore in water depths of 4.3-6.6 m, the middle bar
is 400-500 m offshore and the intervening trough is commonly
more than 10 m deep. The general pattern of bars appears to be
consistent through time but short term variations, similar to
those reported for Kouchibouguac, were recorded between 1974
and 1975. At the more sheltered east coast site, there is a
well-defined linear bar, 200-300 m offshore (Fig. 10C), in water depths which varied from 1.7 m in the summer to 2.5 m in
winter, separated by a distinct trough from a series of small
inner bars, the orientation of which responds very rapidly to
changes in the direction of wave approach.

Offshore Zone

The comments here refer to the shoreface, which is defined
as that section of the barrier island sand body seawards of the
outer nearshore bar to the subtidal-exposure of bedrock or till.
Little attention has been paid to morphology and sediment movement within the zone, but it is clear that there are considerable variations throughout the Southern Gulf. At Malpeque (Armon 1975), there is an overall increase in the width and maximum depth of the shoreface to the southeast or downdrift. In
the north it is 1200-3500 m wide, with maximum depths of 13-17 m;
at Kouchibouguac, bedrock outcrops between 700 and 2000 m offshore at depths between 6 and 9 m (Kranck 1967). Recent offshore surveys along the Miscou-Neguac barriers in New Brunswick
(McCann, in press) indicate that the shoreface sand zone is very
narrow away from inlet locations; bedrock outcrops close to
shore in depths less than 4 m at many locations. From these
data, and similar surveys being undertaken in 1978 in other
areas, it should be possible to make some preliminary assessment of the role of the shoreface in overall barrier island development in the Southern Gulf.

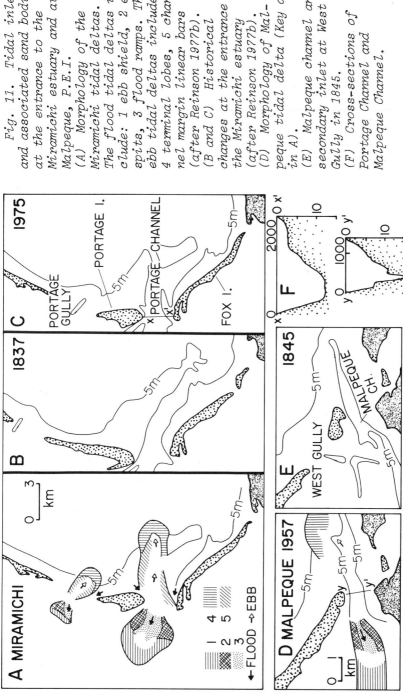

Fig. 11. Tidal inlets and associated sand bodies at the entrance to the Miramichi estuary and at Malpeque, P.E.I.
(A) Morphology of the Miramichi tidal deltas. The flood tidal deltas include: 1 ebb shield, 2 ebb spits, 3 flood ramps. The ebb tidal deltas include: 4 terminal lobes, 5 channel margin linear bars (after Reinson 1977b).
(B and C) Historical changes at the entrance to the Miramichi estuary (after Reinson 1977b).
(D) Morphology of Malpeque tidal delta (Key as in A).
(E) Malpeque channel and secondary inlet at West Gully in 1845.
(F) Cross-sections of Portage Channel and Malpeque Channel.

Tidal Inlets

The tidal inlets of the barrier systems of the Southern
Gulf range in size from very small temporary inlets, only tens
of meters wide and 2-3 m deep, to large permanent inlets over
1 km wide and 15 m deep at the throat section (McCann et al.
1977). Some of the larger inlets have remained in the same position throughout the period of historical record, though there
have usually been considerable changes in the pattern of the
associated channels and shoals. Others have migrated alongshore for distances up to 4 km, but maintained a relatively
constant cross-sectional area during this process.

The largest inlets are Portage Channel at the entrance to
the Miramichi estuary and Malpeque Channel in Prince Edward
Island (Fig. 11). Both exhibit large flood tidal deltas,
which conform well to the model proposed by Hayes (1975), though
both deltas are completely subtidal. Reinson (1977) suggests
that changes in the ebb delta-shoal at Portage Channel during
the past 100 years are related to dredging activities across
the flood tidal delta, which resulted in an increase in the in-
tensity and volume of the ebb discharge such that a channel was
scoured across the shoal. At Malpeque there has been an extension and consolidation of the ebb-tidal delta-shoal complex
since 1845, following the closure of West Gully, a former inlet
some 2 km to the north (Fig. 11).

Two other inlets in the Miramichi and Malpeque systems,
Portage Gully and Hardy's Channel, provide the best examples of
inlets which have been present for over 100 years, but have migrated southwards, downdrift, during this period. Portage
Gully has migrated 4.5 km in the period 1837 to 1975, but has
continued to exhibit symmetrical flood and ebb tidal deltas
during this time. The flood tidal delta is again completely
subtidal (Reinson, 1977). Hardy's Channel has migrated 1200 m
to the southeast since 1765, 350 m of this having occurred since
1935. The flood tidal delta, which is considerably larger than
the ebb tidal delta, is emergent at low tide. The surface of
the delta is 0.4 - 0.7 m above LLW and the sandy body is only
completely inundated at large and spring tides. Major changes
in the channel pattern across this delta appear to influence
conditions at the inlet throat and the rate and direction of
inlet migration (Armon, in press). The morphology and recent
evolution of Hardy's Channel and the small inlet to the north,
documented by Armon, are probably typical of many of the smaller
inlets in the Southern Gulf which are backed by small lagoons
and have small tidal prisms. Both Reinson and Armon stress the
importance of tidal prism as a controlling variable in the development of flood tidal deltas.

Tidal inlets, both present and past, have been the most important locations for the landward transfer of sediment in the

basically transgressive barriers of the Southern Gulf. Thus, the extensive shallow sand area on the landward side of the Miramichi barriers is a region of coalescing flood tidal deltas (Reinson 1977) largely associated with the present inlets. At Malpeque, on the other hand, overwash and aeolian transport at the sites of former inlets constitutes an important part of the landward transfer (Armon, this volume). The lobate character of the lagoonal shoreline at both Malpeque and Kouchibouguac (Bryant and McCann, 1973) reflects the frequency and importance of inlet breaching. Initially, the lobes were small flood tidal deltas at temporary inlets, which received additional sediment by overwash following inlet closure.

CONCLUSIONS

(1) This account has stressed the variability of barrier shoreline conditions within a small enclosed sea, the Southern Gulf of St. Lawrence, characterized by low tidal ranges and short period, locally-generated waves. Tidal conditions are relatively uniform throughout, but wave energy levels differ in the four principal barrier systems, according to shoreline orientation in relation to available fetches.

(2) The prevailing winds are westerly and the most dynamic beaches occur on the west coast of the Magdalen Islands in the center of the Gulf. The New Brunswick and Prince Edward Island barriers are on "lee" shores with respect to the prevailing winds, and the important waves come from the northeast quadrant.

(3) The uniformly low tidal ranges do not impart a typically microtidal (as opposed to mesotidal) condition to the barriers. The diagnostic features of both microtidal and mesotidal barriers, outlined by Hayes (1976, Table 7, p. 1-96), occur within the Southern Gulf.

(4) In particular, tidal inlets and tidal deltas, which Hayes considers to be of relatively minor significance in microtidal barriers, are important in the Southern Gulf. They range from large inlets, which have been present throughout the period of historical record (200 years) to small temporary inlets, which may last for only five years. The largest inlets and associated sand bodies occur at the southern, or downdrift, ends of the three "lee" shore barrier systems, opposite the largest drowned estuaries. The temporary inlets become the sites of overwash, following inlet closure.

(5) Away from the main tidal inlets the variation in barrier morphology around the Southern Gulf is best exemplified by the contrast between the high dune, restricted overwash condition at Malpeque, and the low overwashed barriers at

Tabusintac (cf. Armon and Rosen, this volume). Both sections
of shoreline are undergoing landward retreat at present, but
appear to be maintaining a relatively constant form and volume
of supratidal sediments in the process.

(6) The variations in wave energy throughout the Gulf result in variations in foreshore and nearshore morphology. Thus,
the system of nearshore bars is wider and extends seawards to
greater depths on the exposed west coast of the Magdalen
Islands than in the protected situation of Kouchibouguac Bay.
The west coast of the Magdalen Islands is also characterized by
a marked seasonal variation in wave energy levels, which produces a distinct "summer-winter" beach cycle (Owens 1977).

(7) Availability of sediment appears to be an important
factor contributing to the variations in barrier shoreline conditions, though information is not complete on this point. The
"downdrift" increase in both the subtidal and supratidal dimensions of the New Brunswick and Prince Edward Island barrier
systems is well documented, and it is tempting to suggest that
the low, narrow barrier condition at Tabusintac (see 5 above)
is due to scarcity of sediment.

(8) Most of the barrier shorelines are narrow and transgressive in character, but little is known of the vertical
sedimentary sequence across the islands. It would seem that
the stratigraphic models developed by Kraft (1971; Kraft et al.
1973) for the Delaware coast would be generally appropriate,
with the addition of various inlet sequences (Hayes 1976; Kumar
and Sanders, 1974); in particular, shallow sequences topped by
overwash deposits, representing the sites of temporary inlets.

ACKNOWLEDGMENTS

This review owes much to discussions and collaboration over
the past years with several of the people whose work is cited
below; in particular John Armon, Ted Bryant, Ed Owens and Gerry
Reinson. The manuscript was prepared while I was on the staff
of the Atlantic Geoscience Centre of the Geological Survey of
Canada at the Bedford Institute of Oceanography.

REFERENCES

Armon, J.W. (1975). The dynamics of a barrier island chain,
    Prince Edward Island, Canada. Ph.D. thesis, McMaster Univ.,
    Hamilton, Ont., 546 p.
Armon, J.W. (this volume). Landward sediment transfers in a
    transgressive barrier island system, Canada.

Armon, J.W. (in press. Changeability in small flood tidal del-
tas and its effects, Malpeque barrier system, Prince Edward
Island. *In* "The Coastline of Canada" (S.B. McCann, ed.),
Geological Survey of Canada.

Armon, J.W. and McCann, S.B. (1977a). The establishment of an
inshore wave climate and longshore sediment transport rates
from hourly wind data. Discussion Paper No. 9, Monograph
Ser., Dept. of Geography, McMaster Univ., 76 p.

Armon, J.W. and McCann, S.B. (1977b). Longshore sediment transport and a sediment budget for the Malpeque barrier system,
Southern Gulf of St. Lawrence. *Can. J. Earth Sci. 14*, 2429.

Armon, J.W. and McCann, S.B. (in press). Morphology and landward sediment transfer in a transgressive barrier island
system, Southern Gulf of St. Lawrence. *Marine Geology*.

Ashe, G.W.T. and Ploeg, J. (1971). Wave climate study, Great
Lakes and Gulf of St. Lawrence. Mech. Eng. Report MH-107A,
vol. 2, N.R.C., Ottawa, Canada, 258 p.

Atmospheric Environment Service (1972). Summary of synoptic
meteorological observations. Atmospheric Environment Service, 4905 Dufferin St., Toronto, Ontario.

Bartlett, G.A. and Molinsky, L. (1972). Foraminifera and Holocene history of the Gulf of St. Lawrence. *Can. J. Earth
Science 9*, 1204-1215.

Bryant, E.A. (1972). The barrier islands of Kouchibouguac Bay,
New Brunswick. M. Sc. Thesis, McMaster Univ., 277 p.

Bryant, E.A. (1974). A comparison of air photography and com-
puter simulated wave refraction patterns in the nearshore
area, Richibucto, Canada, and Jervis Bay, Australia. *Maritime Sediments 10*, 85-95.

Bryant, E.A. and McCann, S.B. (1972). A note on wind and wave
conditions in the Southern Gulf of St. Lawrence. *Maritime
Sediments 8*, 101-103.

Bryant, E.A. and McCann, S.B. (1973). Long and short term
changes in the barrier islands of Kouchibouguac Bay, Southern Gulf of St. Lawrence, Canada. *Can. J. Earth Sci. 10*,
1582-1590.

Davidson-Arnott, R.A.D. (1975). Form, movement and sedimentological characteristics of wave formed bars: a study of their
role in the nearshore equilibrium, Kouchibouguac Bay, New
Brunswick. Ph.D. thesis, Univ. of Toronto.

Davidson-Arnott, R.G.D. and Greenwood, B. (1974). Bedforms and
structures associated with bar topography in the shallow
water wave environment, Kouchibouguac Bay, New Brunswick,
Canada. *J. Sed. Petrology 44*, 698-704.

Davidson-Arnott, R.G.D. and Greenwood, B. (1976). Facies rela-
tionships on a barred coast, Kouchibouguac Bay, N.B., Can.
In "Beach and Nearshore Sedimentation" (R.A. Davies and R.L.
Ethington, eds.), 149-168. Special Publ. No. 24, Soc. Econ.
Paleon. and Mineralogists, Tulsa, Oklahoma.

Dolan, R. (1971). Coastal landforms: crescentic and rhythmic. *Geol. Soc. Amer. Bull. 82*, 177-180.

Farquharson, W.I. (1959). Causeway investigation Northumberland Strait: Report on tidal survey 1958. Can. Dept. Energy, Mines and Resources, Surveys and Mapping Branch, 137 p.

Farquharson, W.I. (1968). Gulf of St. Lawrence tides and tidal streams. *In* "Gulf of St. Lawrence Workshop" (R.W. Treites, ed.), p. 8-23. Bedford Institute of Oceanography.

Forward, C.N. (1954). Ice distribution in the Gulf of St. Lawrence during the breakup season. *Geog. Bull. 6*, 45-84.

Frankel, L. and Crowl, G.H. (1961). Drowned forests along the eastern coast of Prince Edward Island, Can. *J. Geol. 69*, 352.

Ganong, W.F. (1908). The physical geography of the north shore sand islands. *Bull. of the N.B. Nat. His. Soc., no. 26, 6*, 22-29.

Grandtner, M.M. (1966). Premières observations phytopedologique sur les prés salés des Iles-de-la-Madeleine. *Le Natur. Can. g 3*, 361-366.

Grandtner, M.M. (1968). Quelques observations sur la végétation psammophile des Iles-de-la-Madeleine. Collectanea Botanica, VII, 25, 519-530.

Grandtner, M.M. (1970). Ecological study of the interior dunes of West Brockley Beach, Prince Edward Island. Preliminary Report to National Parks Branch, Dept. of Indian and Northern Affairs, Canada.

Grant, D.R. (1970). Recent coastal submergence to the Marine Provinces, Canada. *Can. J. Earth Sci. 7*, 676-689.

Greenwood, B. and Davison-Arnott, R.G.D. (1972). Textural variations in the subenvironments of the shallow water wave zone, Kouchibouguac Bay, N.B. *Can. J. Earth Sci. 9*, 679-688.

Greenwood, B. and Davidson-Arnott, R.G.D. (1975). Marine bars and nearshore sedimentary processes. *In* "Nearshore sediment dynamics and sedimentation" (J. Hails and A. Carr, eds.), p. 123-150. John Wiley and Sons, New York.

Greenwood, B. and Hale, P. (in press). Depth of activity, sediment flux and morphological change in a barred nearshore environment. *In* "The Coastline of Canada" (S.B. McCann, ed.). Geol. Surv. Can.

Hale, P. and Greenwood, B. (in press). Storm wave climatology: a study of the magnitude and frequency of geomorphic process. *In* "The Coastline of Canada" (S.B. McCann, ed.). Geol. Surv. Can.

Hayes, M.O. (1975). Morphology of sand accumulation at inlets: an introduction to the symposium. *In* "Estuarine Research" Vol. II (L.E. Cronin, ed.), p. 3. Academic Press, New York.

Hayes, M.O. (1976). Transitional Coastal Environments. *In* "Terrigenous clastic depositional environments" (M.O. Hayes and T. Kana, eds.). T.R. 11-CRD, Dept. Geol., Univ. S. Carolina.

Hayes,M.O. and Boothroyd, J.C. (1972). Comparison of ridge and runnel systems in tidal and non-tidal environments. *J. Sed. Petrology 42*, 413-421.

Huntley, D.A. (in press). Edge waves in a crescentic bar system. *In* "The Coastline of Canada" (S.B. McCann, ed.). Geol. Surv. Can.

Johnson, D.W. (1925)."The New England Acadian shoreline," 608 p. Wiley and Sons, N.Y. (Facsimile Edition: Hafner, N.Y.'67).

Komar, P.D. (1971). Nearshore cell circulation and the formation of giant cusps. *Geol. Soc. Amer. Bull. 82*, 2643-2650.

Kraft, J.C. (1971). Sedimentary facies pattern and geologic history of a Holocene marine transgression. *Geol.Soc. Amer. Bull. 82*, 2131-2158.

Kraft, J.C., Biggs, R.B., and Halsey, S.D. (1973). Morphology and vertical sedimentary sequence models in Holocene transgressive barrier systems. *In* "Coastal Geomorphology" (D.R. Coates, ed.). Publ. in Geomorph., SUNY, Binghamton.

Kranck, K. (1967). Bedrock and sediments of Kouchibouguac Bay, N.B. *J. of Fisheries Res. Board of Can. 24*, 2241-2265.

Kranck, K. (1972a). Geomorphological developments and Post-Pleistocene sea-level changes, Northumberland Strait, Maritime Provinces. *Can. Earth Sci. 9*, 835-844.

Kranck, K. (1972b). Tidal current control of sediment distribution in Northumberland Strait, Maritime Provinces, Canada. *J. Sed. Petrology 42*, 596-601.

Lamoureux, G. and Grandtner, M.M. (1977). Contribution à l'étude ecologique des dunes mobiles. Les éléments phytosociologiques. *Can. J. Bot. 55*, 158-171.

Loring, D.H. (1973). Marine geology of the Gulf of St. Lawrence. Geol Surv. Can., Paper 71-23, 305-324.

Loring, D.H. and Nota, D.J.G. 1969. Mineral dispersal patterns in the Gulf of St. Lawrence. *Rev. Geog. Montreal 23*, 289-305.

Loring, D.H. and Nota, D.J.G. (1973). Morphology and sediments of the Gulf of St. Lawrence. Fisheries Res. Board of Can. Bull. #182, 147 p.

Loring, D.H., Nota, D.J.G., Chesterman, W.E., and Wong, H.K. (1970). Sedimentary environments on the Magdalen Shelf, Southern Gulf of St. Lawrence. *Marine Geol. 8*, 337-354.

Matheson, K.M. 1967. The meteorological effect on ice in the Gulf of St. Lawrence. McGill Univ. Marine Science Centre, Unpubl. Rept. #3, 110 p.

McCann, S.B. (1972). Reconnaissance Survey of Hog Island, Prince Edward Island. *Maritime Sediments 8*, 107-113.

McCann, S.B. (in press). Offshore surveys along the barrier island shorelines of NE New Brunswick and N. Prince Edward Island. Current Research, Geol. Surv. Canada.

McCann, S.B. and Bryant, E.A. (1970). Beach processes and shoreline changes, Kouchibouguac Bay, New Brunswick. *Maritime Sediments 6*, 116-117.

McCann, S.B. and Bryant, E.A. (1972). Barrier islands, sand spits and dunes in the Southern Gulf of St. Lawrence. *Maritime Sediments 8*, 104-106.

McCann, S.B. and Bryant, E.A. (1973). Beach changes and wave conditions, New Brunswick. Proc. 13th Conf. Coastal Eng., Vancouver, Vol. II, 1293-1304.

McCann, S.B., Bryant, E.A., and Seeley, R.S. (1973). Barrier island, shoreline and dune survey, Kouchibouguac National Park. Report to National Parks Branch, Dept. Indian and Northern Affairs, Canada, 163 p.

McCann, S.B., Reinson, G.E., and Armon, J.W. (1977). Tidal inlets of the Southern Gulf of St. Lawrence, Canada. Coastal Sediments '77, Proc. Symp. Coast. Sed. & Struct.(ASCE), 504.

Miramichi Channel Study (1978). Draft Final Report. Marine Directorate, Dept. of Public Works, Canada.

Munroe, H.D. (1976). The effect of storms on nearshore morphology, Neguac Island, New Brunswick. Geol. Surv. Canada, Paper 76-1C, 37-39.

Munroe, H.D. (1977). Historical development of Neguac Island, New Brunswick, Canada. Proc. 4th Int. Conf. Port and Ocean. Eng. under Arctic Conditions, St. John's, Newfoundland, 916.

Owens, E.H. (1974a). Coastline changes in the Southern Gulf of St. Lawrence. Geol. Surv. Can., Paper 74-1A, 123-124.

Owens, E.H. (1974b). A framework for the definition of coastal environments in the Southern Gulf of St. Lawrence. *In* "Offshore Geology of Eastern Canada" Vol. 1 (B.R. Pelletier, ed.), p. 47-76. Geol. Surv. of Can., Paper 74-30.

Owens, E.H. (1975a) The geodynamics of two beach units in the Magdalen Islands, Quebec, within the framework of coastal environments of the Southern Gulf of St. Lawrence. Ph.D. thesis, Univ. of South Carolina, 319 p.

Owens, E.H. (1975b). Barrier beaches and sediment transport in the Southern Gulf of St. Lawrence. Proc. 14th Coastal Eng. Conf. (ASCE, N.Y.), Vol. II, 1177-1193.

Owens, E.H. (1976). The effects of ice on the littoral zone at Richibucto Head, eastern New Brunswick. *Rev. Geog. Montreal 30*, 95-104.

Owens, E.H. (1977a). Temporal variations in beach and nearshore dynamics. *J. Sed. Petrology 47*, 168-190.

Owens, E.H. (1977b). Process and morphology characteristics of two barrier beaches in the Magdalen Islands, Gulf of St. Lawrence, Canada. Proc. 15th Conf. Coast. Eng., Honolulu, 1975-1991.

Owens, E.H. and Frobel, D. (1977). Ridge and runnel systems in the Magdalen Islands, Quebec. *J. Sed. Petrology 47*, 191-198.

Owens, E.H. and Harper, J. (1972). The coastal geomorphology of the Southern Gulf of St. Lawrence: a reconnaissance. *Maritime Sediments 8*, 61-64.

Owens, E.H. and McCann, S.B. (in press). The coastal geomorphology of the Magdalen Islands. Quebec Geol. Surv. Can.

Ploeg, J. (1971). Wave climate study, Great Lakes and Gulf of St. Lawrence. Mech. Eng. Rept. MH-107A, Vol. 1, N.R.C., Ottawa, Canada, 160 p.

Prest, V.K., Terasmae, J., Mathews, J.V., and Lichti-Federovich, S. (1976). Late-Quaternary history of Magdalen Islands, Que. *Maritime Sediments 12*, 39-58.

Quon, C., Keyte, F.K., and Pearson, A. (1963). Comparison of five years hindcast wave statistics in the Gulf of St. Lawrence and Lake Superior. Bedford Inst. Ocean., Rept. 63-2, 59 p.

Reinson, G.E. (1976a). Channel and shoal morphology in the entrance to the Miramichi estuary. Geol Surv. Can., Paper 76-1C, 33-35.

Reinson, G.E. (1976b). Surficial sediment distribution in the Miramichi estuary, New Brunswick. Geol. Surv. Can., Paper 76-1C, 41-44.

Reinson, G.E. (1977a). Examination of bedforms in shallow water using side scan sonar, Miramichi estuary, New Brunswick. Geol. Surv. Can., Paper 77-1B, 99-105.

Reinson, G.E. (1977b). Tidal current control of submarine morphology at the mouth of the Miramichi estuary, N.B. *Can. J. Earth Sci. 14*, 2524-2532.

Reinson, G.E. (in press). Variations in tidal inlet morphology and stability, eastern New Brunswick. *In* "The Coastline of Canada" (S.B. McCann, ed.). Geol. Surv. Can.

Rosen, P. (this volume). Aeolian dynamics of a barrier island system.

Thomas, M.L.H., Grant, D.R., and deGrace, M. (1973). A late Pleistocene shell deposit at Shippegan, New Brunsick. *Can. J. Earth Sci. 10*, 1329-1332.

Vanicek, P. (1976). Pattern of recent vertical crustal movements in Maritime Canada. *Can. J. Earth Sci. 13*, 661-667.

Willis, D.H. (1977). Evaluation of alongshore transport models. Coastal Sediments '77, Proc. Symp. on Coast. Sed. and Struct. (ASCE), 350-365.

Dohler, G. (1969). Tides in Canadian waters. Can. Hydrogr. Service, 14 p.

LANDWARD SEDIMENT TRANSFERS IN A TRANSGRESSIVE BARRIER ISLAND SYSTEM, CANADA

John W. Armon

Department of Geology
Memphis State University
Memphis, Tennessee

The largely transgressive Malpeque barrier system in the southern Gulf of St. Lawrence, Canada, was investigated to assess the influence of a restricted sea environment on barrier island morphology and dynamics. Wave dimensions are restricted by the limited wind fetches and shallow water depths in the southern Gulf, and the islands are protected from storm action for 2-4 months in winter by shorefast ice. The barrier islands in the transgressive section of this barrier system are narrow, 150-450 m in width, with the dune zone commonly dominated by the seawardmost dune. The continuous dune cover restricts overwashing during storms, and the well-vegetated dune surface limits sand transport landward of the seawardmost dune. Consequently, over 90% of measured sediment gains to the barrier system during the 33 year period, 1935-1968, were located in the vicinities of existing or closed inlets. Of these inlet-associated gains, over 40% by volume occurred at the sites of "temporary" inlets, open for only part of the 210 year period of map record. This reflected dune and foreshore recovery at former inlet sites, resulting largely from wind action and storm overwash. In all, 48% of the measured volumetric gains to the barrier system arose from storm overwash and wind transport at locations other than contemporary inlets, while the remaining 52% were measured at the open inlets. Comparisons drawn between the transgressive barrier coast at Malpeque and analogous coasts to the south suggest some variability in the dynamic response among barriers in different coastal environments.

65

INTRODUCTION

This paper reports on a study of the largely transgressive Malpeque barrier system, located in the southern Gulf of St. Lawrence, Canada (Fig. 1). The coastal environment in the southern Gulf is different in some respects to that along the Atlantic seaboard of the United States, where many investigations of barrier island responses have been undertaken. Consequently, the examination of shoreline morphology and changes along the Malpeque barrier coast presents an opportunity to assess the influence of a restricted sea environment on transgressive barrier systems. The information presented in this report is complemented by that of Rosen (this volume) from a contrasting barrier shoreline in the southern Gulf of St. Lawrence. The regional differences around the southern Gulf are described and discussed more fully by McCann (this volume). Barrier islands maintain their general dimensions during transgression by the actions that transport sand landward from the littoral zone--notably wind, storm overwash, and currents at tidal inlets (Fig. 2). This paper focuses on the relative significances of these processes at Malpeque, as established from volumetric measurements of barrier island and inlet changes over a 33-year period (Armon 1975; Armon and McCann, in press).

Numerous observations have been made regarding landward transport by wind and storm overwash for barrier systems on open ocean coasts. Dillon (1970, p. 104) was perhaps the first person to consider the problem conceptually, and he reported for the Charleston Pond barrier, Rhode Island, that "Sand is supplied to the barrier back almost exclusively by washover and wind transport from the barrier front...." Cross-sections of the low barrier islands west of Cape Hatteras, North Carolina (Godfrey and Godfrey, 1973), record details of island buildup and migration resulting from storm overwash and wind transport. Such evidence led Dolan (1973, p. 263) to conclude that "Overwash is the only means by which massive quantities of coarse sediment can be moved inland...." Observations by other workers also suggest that overwash is a significant action on some barriers (Moody 1964; Shepard and Wanless, 1971; Kraft 1971). In contrast, recent investigations elsewhere raise questions about the general effectiveness of overwash in transporting sand landward, in view of the subsequent movement of the overwashed sand by wind (Schwartz 1975; Leatherman 1976; Fisher and Stauble, 1977). Offshore winds may return the overwashed sand to the foreshore during the late or post-storm recovery. This redistribution by wind often means that the two actions cannot be separated and have to be considered together. Along some barrier coasts aeolian sand transport landward is confirmed by the extensive dune systems present. However, sand

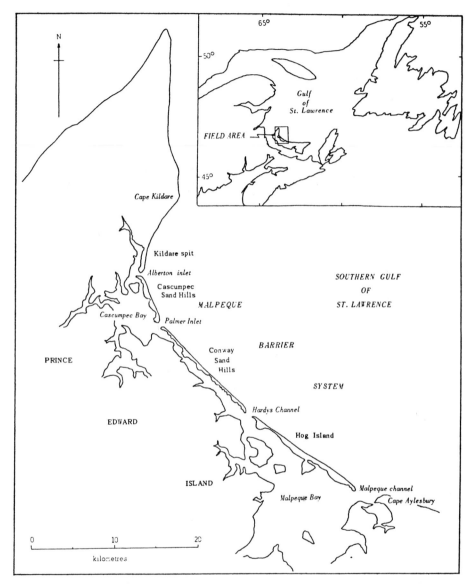

Fig. 1. *Malpeque barrier system, southern Gulf of St. Lawrence, Canada.*

transport in the dune zone is impeded by vegetation, so that contemporary aeolian transport is most important on vegetation-free or open surfaces, where onshore winds prevail.

   Tidal inlets have long been recognized as major sites for landward sand transport from the littoral zone, by current,

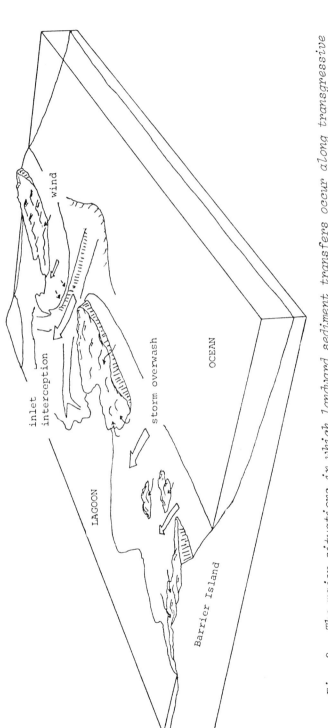

Fig. 2. The major situations in which landward sediment transfers occur along transgressive barrier shorelines.

wave, and wind actions. Caldwell (1966) considered all of the
sand eroded from the southern New Jersey coast to be lost landward through seven tidal inlets. Results from other barriers
confirm that large volumetric totals are intercepted at some
tidal inlets (Moody 1964; McCormick 1973). In addition, data
from two other studies indicate that inlets are the sites for
the major proportion of total sand movements in barrier systems. Volumetric subtotals from Pierce (1969) for the barrier
islands west of Cape Hatteras suggest that 70% of the landward
sediment transfer has been taking place at inlet sites. (This
neglects the accretion on Cape Lookout and its subtidal shoal.)
Information presented in Bartberger (1976) indicates that 82%
of the sand transfer on Assateague Island in the last 5000
years has been at tidal inlets, with the remainder taking place
by storm overwash and wind transport.

The volumetric data suggest that tidal inlets play the
dominant role in moving sand landward and maintaining barrier
systems during transgression. The other actions appear to
transport relatively small proportions of sand landward. However, overwash and wind action cannot be readily discounted for
some barrier systems. These actions proceed adjacent to inlet
margins, hence contributing to some of the inlet transfer. Also,
following inlet closure the resulting low open surfaces provide
sites for additional landward sand movements by wind and storm
overwash.

Information from the Malpeque barrier system has enabled
the calculation of landward sediment transport at tidal inlets
and other locations. Thus, the significance of tidal inlets
versus other transport mechanisms along the Malpeque coast
can be gauged. The contrasting coastal environments between
the southern Gulf and the Atlantic seaboard of the United
States also mean that an assessment can be made of the impact
of environmental differences in terms of barrier dynamics.

MALPEQUE BARRIER SYSTEM

The Malpeque barrier system is located on the north shore of
Prince Edward Island, with a wind fetch of more than 500 km to
the northeast. Lesser wind fetches exist to the north and east.
However, shallow water depths of 40-60 m in the southern Gulf
probably limit wave dimensions more than the wind fetches.
Maximum recorded significant wave heights in "deep water" ap-
proach 5 m, and breaker heights of 2-3 m have been observed at
Malpeque during summer storms. Consequently, the coastal

environment cannot be considered a low energy one. Calculations
using procedures outlined in U.S. Army C.E.R.C. report (1966)
suggest that storm surges of 1 m are likely for the slopes
present seaward of the barrier system during moderate to extreme storms. However, water level records or direct field
observations are lacking for this shore.

The most significant difference between the coastal environment here and that along the ocean coasts to the south is
introduced by the presence of ice in the coastal zone for 2-4
months each winter. Shorefast ice approaches thicknesses of
60 cm and commonly occurs for 1-2 km seaward from the shore.
It provides protection for the barrier system during the most
stormy period of the year, thus decreasing the frequency of
coastal damage by extreme storms. The dunes also remain frozen
for much of this period, with the result that wind damage of
the dune system during storms is also minimized. This information suggests that extreme storm damage is likely to be encountered less frequently along the Malpeque barrier chain than
on more open ocean coasts. The tides in the southern Gulf also
have lower tidal ranges than those along the Atlantic coast.
The tides at Malpeque are mixed mainly semidiurnal tides, with
a microtidal range (spring tides up to 1.3 m).

The Malpeque barrier system is a 43 km long sandy barrier
shoreline, in a natural condition, with three islands and a
spit. It consists of a transgressive northwestern section and
a progradational southeastern section. Northwest of central
Hog Island, the morphology of the barrier islands is largely
related to coastal actions in the last 200 years--inlet development or closure, localized overwash, dune building, and
general shoreline retreat. In the southeast, most of Hog
Island includes a dune ridge sequence resulting from shoreline
progradation at some time within the last 1500-2500 years.
This area is omitted from further consideration in this paper
because of its contrasting development. Rates of coastal erosion established from aerial photography are low, with an average of 0.26 m year^{-1} for the entire barrier shoreline.

The islands characteristically have an irregular, lobate,
lagoonal margin with island widths varying between 150 and
450 m (Fig. 3). The variability is normal but the minimum
widths commonly encountered at Malpeque (150-200 m) are less
than those usually present on open ocean barriers. This appears to be a response to erosion of the foreshore and dune
margin in situations where little sand is moved directly landward by wind or storm overwash. Such a condition is common at
Malpeque because of the well-vegetated continuous dune cover
generally present landward of the foreshore (Fig. 4) and the
dune elevations of 3-5 m. The only breaks in the dune system
occur at the tidal inlets, at a washover on Hog Island, and at
some small washover channels in the northwest (Fig. 3). The

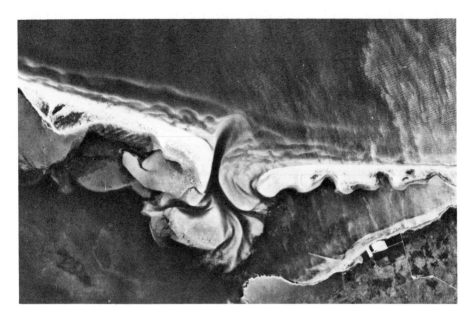

Fig. 3. Barrier island planform - southern Cascumpec Sand
Hills, Palmer Inlet, and Conway Sand Hills, 1964.
Source: A18452-197, National Air Photo Library, Ottawa, Canada.

Fig. 4. Vegetation cover in the dune zone on western Hog
Island. View southeast.

well-vegetated continuous dune condition is thought to arise as
a result of the low frequency of devastation by extreme storms
along the Malpeque coast. General island overwash has not occurred since at least 1935 and there is no indication in the
present morphology that it was ever widespread. In the general
absence of island overwash or significant landward aeolian
transport, shoreline recession produces island sections with
narrow widths.

Narrow areas in the barrier island are favourable for breaching by wave attack from the seaward side, with inlet formation
commonly resulting. At least nine inlets in addition to the
present ones have occurred along the Malpeque shore during the
last 200 years. Eight of these were located in the western
transgressive section. The occurrence of such "temporary" inlets, now filled, has done much to create the lobate lagoonal
margin of the barrier islands.

LANDWARD SEDIMENT TRANSFER AT MALPEQUE

At Malpeque an attempt was made to measure the volumes of
all landward sand movements in the transgressive section during
a 33-year period covered by aerial photography, 1935-1968.
Zones undergoing change in this period were mapped from the
photographs and areal changes measured. The mean thicknesses
of areas undergoing change were calculated by averaging surface
heights at 1 m intervals from representative lines surveyed
across the foreshore or dune sections in 1972-1974. Two or
more profile lines were surveyed where there was substantial
variability in the surface within an area. The product of area
and mean thickness for any location gave the volumetric change.

The changes measured include dune establishment on surfaces
open in 1935, foreshore recovery, and subtidal or subaerial accretion at inlet margins and on flood tidal deltas. The thicknesses of accumulations mapped around the margins of the flood
tidal deltas were established from knowledge of former lagoon
depths and elevations of the depositional surface. In zones
where accretion of the flood tidal delta surface was evident,
thicknesses were established by comparing the 1935 and 1968
photographs, both of which were recorded close to low tide.
Estimations were then made on the basis of tonal differences
across the low intertidal and shallow subtidal locations.

The procedure just outlined is possible in this barrier
system because of certain favourable conditions. Overwash is
limited by the continuous dune cover to tidal inlet margins and
a few other locations. In addition, the dunes are well vegetated landward of the foreshore and sand movements within the
vegetated zone can be neglected. Thus, dune accretion is

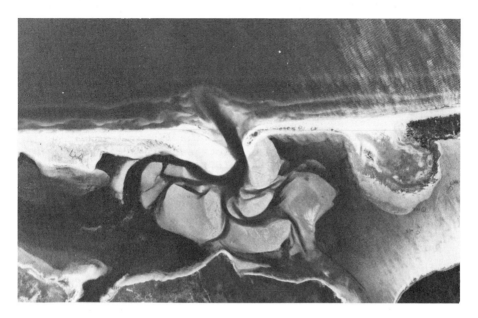

Fig. 5.   Hardys Channel, 1964, and the washover on western Hog Island.   Source: A18452-192, National Air Photo Library, Ottawa, Canada.

Fig. 6.   Site of Cavendish Inlet, closed in 1938.   Dune recovery is almost complete, with only small washover channels remaining open by 1974.

generally limited to zones undergoing dune formation between
1935 and 1968. Elsewhere, dune accretion is only likely on
the seawardmost dune. Such accretion was not measured but the
volumes are probably minor. This sand often represents an internal transfer from the seaward to the landward slope of the
dune; it can be removed by subsequent erosion of the dune, and
two years' survey results indicate minor changes in general on
the seaward dune.

The results as presented in Table I are considered as annual volumetric sand movements from the littoral zone. Distinc-
tions have been drawn between inlet-associated sand movements,
those at former inlet sites resulting from overwash and wind
action, and "other" sand transport. The nature of the measurements made at Malpeque means that sand volumes transported by
storm overwash and wind have to be treated together.

Approximately 47100 m^3 $year^{-1}$, or 52% of the measured landward sand transfer, occurred at inlet sites during the survey
period. Most of this amount reflected accretion on the flood
tidal deltas and inlet margins of two inlets, Hardys Channel
(Fig. 5) and Palmer Inlet (Fig. 3). A significant portion of
the sand deposited at these inlets was transported by storm
overwash and wind. Dune and foreshore accretion was prominent
on the updrift margin of Hardys Channel, accompanying its
southeastward migration during the survey period. At Palmer
Inlet, both inlet margins underwent changes associated with
foreshore development and landward movement of the waterlines.
Interception by tidal currents appears to account for 70-80% of
the landward sand transfer at these inlets.

The remaining 48% of the measured landward sand transfer resulted from overwash and wind action operating elsewhere along
the barrier shore. Most of this 44000 m^3 $year^{-1}$ was recorded
at the sites of five former inlets, present at some time in the
last 210 years. Thus temporary tidal inlets on this coast are
important for ensuring the continued landward transport by
storm overwash and wind. Even small inlets produce open areas
upon closure and provide sites for this additional sand movement until dune and waterline recovery is complete. Cavendish
Inlet, open between 1920 and 1938, underwent dune establishment
into the 1970s (Fig. 6). By 1974 overwash was restricted to a
few "channels," and some dunes were between 1 and 3 m in height.
The large washover on western Hog Island (Fig. 5) has remained
vegetation-free since closure of the original inlet in the late
19th century, but dunes are steadily encroaching onto the fan
surface. Observations on this and other washovers confirm that
wind redistributes most of the overwashed sand, but the growth
of dunes 2-5 m high at washover margins indicates that a substantial proportion of this sand is moved laterally.

Accretion from overwash and wind action in sections other
than inlet locations was only a small proportion of the volumes

TABLE I.   *Landward Sediment Transfers within the Transgressive Section of the Malpeque Barrier System, 1935-1968.*

Sediment Transfer	Volume (m^3yr^{-1})	Subtotals
INLET-ASSOCIATED SAND TRANSFER		
1. Accretion on flood-tidal deltas	24,600	
2. Alongshore migration of inlet margins	16,000	*(overwash, wind & interception)*
3. Landward migration of inlet margins	6,500	47,100
SAND TRANSFER AT FORMER INLET SITES		
Dune and waterline recovery by wind and overwash	38,300	
OTHER MEASURED SAND TRANSFER		*(overwash and wind action)*
Dune recovery in overwash features	5,700	
		44,000
TOTAL MEASURED MOVEMENTS		91,100

measured in this 33-year period.   The amount recorded was for a minor washover on western Conway Sand Hills, which progressively filled during the decades following 1935.

DISCUSSION AND CONCLUSIONS

The information presented for the Malpeque barrier system indicates that the landward sediment transfer at tidal inlets for the 1935-1968 period was substantially less significant than estimated by Pierce (1969) or Bartberger (1976).   At Malpeque the proportion moved by storm overwash and wind was approximately 60% of the totals measured.   The differences between the Malpeque barrier system and those to the south appear to be a response to the relatively small volumes moved landward at tidal inlets along the Malpeque coast.   Measured accretion suggests that only 1.4 m^3 year^{-1}/m shore were moved landward at inlets along the Malpeque shore compared with the estimated

3.8 m^3 year^{-1}/m shore west of Cape Hatteras and 6.5 m^3 year^{-1}/m shore for Assateague Island (Table II). Amounts moved by overwash and wind are similar for all three areas, however. The small standardized sediment volumes accumulating at inlet sites along the Malpeque shore appear to be a response to the sizes of tidal inlets present. Their small dimensions are a result of minor tidal prisms, occurring because of the microtidal regime and the small lagoon areas backing much of the coast. In addition, landward sand movements at Alberton Inlet are probably limited by its shallow bedrock channel bottom.

In spite of this, inlets play the dominant role in promoting landward sediment movements in the Malpeque barrier system. The data indicate that over 90% by volume of the total landward movements measured there were related to actions at open or closed inlet sites (Table I). This appears to be a response to the low frequency of devastation by extreme storms and the resulting well-vegetated continuous dune cover. Tidal inlets have also produced the wider zones of barrier island in the transgressive section, as flood tidal deltas are partly incorporated into the islands following inlet closure. In all, the morphology of 35% of this part of the Malpeque barrier system has been influenced to some degree by tidal inlets present within the last 210 years.

The information presented here indicates some variability in the significance of landward sand transfers at tidal inlets between barriers from different geographic locations. A question arises concerning the range in importance of tidal inlet transfers that can be expected amongst barrier systems in different situations. This is best considered by looking at the major factors that affect landward sediment transport (Fig. 7). There appear to be two partly independent sets of factors influencing sediment transfers at inlets and at other locations by wind and storm overwash. Tidal and lagoon characteristics influence inlet dimensions and tidal flow through inlets. Waves and currents near inlets affect inlet migration and closure. Also, extreme storm actions can cause inlet formation as well as inlet closure. The resulting changes in the number of inlets present may influence general flow conditions at all inlets. Landward sediment transport by wind and overwash is also influenced by storm actions, as they affect the dune cover, storm overwash and wind transport. Other important factors include vegetation and moisture conditions; sediment abundance in the shore zone influences beach geometry (width, height, volume of sediment) and consequently storm effects.

There is thus a degree of independence between some of the factors influencing transport at inlets and at other locations. Because of this and the contrasts existing in coastal environments, considerable differences are possible in the relative

TABLE II. Landward Sediment Transfers in Different Environments.

Location & Sediment Transfer	Volume, $m^3 year^{-1}$	Percentage	Volume/m shore, $m^3 \ year^{-1}/m \ shore$
CAPE HATTERAS (PIERCE 1969)*			
Tidal Inlets	382,000	72.4	3.8
Overwash	75,000	14.2	0.8
Wind	71,000	13.3	0.7
ASSATEAGUE ISLAND (BARTBERGER 1976)**			
Tidal Inlets	375,000	82	6.5
Overwash	57,000	12	1.0
Wind	30,000	6	0.5
MALPEQUE BARRIER SYSTEM			
Tidal Inlets	47,000	52	1.4
Overwash and Wind	44,000	48	1.3

* averaged over 100 km of shore
** averaged over 58 km of shore

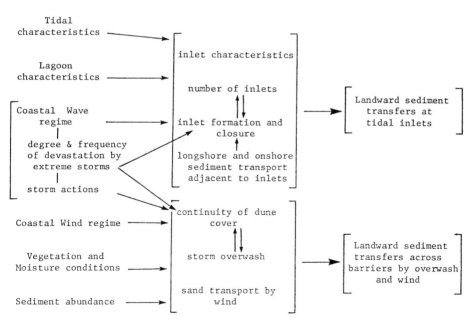

Fig. 7. The major factors influencing the significance of landward sediment transfers at tidal inlets and in other locations by overwash and wind.

significances of tidal currents, storm overwash, and wind action amongst barrier systems. The results from Malpeque and the two barrier coasts to the south may represent the range of variability that can generally be expected. There are large differences in the proportions of sediment transfers at inlets (Table II, volumes/m shore), probably arising from the contrasts in tidal and lagoon characteristics. The coastal responses along other barrier shorelines, however, might give results falling outside this range. At one extreme, slowly migrating barriers with large losses landward at tidal inlets could have closer to 90-95% of sand movements occurring at inlets. Caldwell's (1966) discussion implies 100% of sand transfers at tidal inlets in southern New Jersey. At the other extreme, an absence of tidal inlets in baymouth barriers, or tidal inlets which are present but not intercepting sediment (e.g., Byrne et al. 1975), would imply that the majority of sand movements are a result of overwash and wind transport, at least in the short term.

At Malpeque, low tidal ranges and small lagoons have led to the small landward sand transfers through inlets during the period considered. As a result, landward transport by storm overwash and wind is relatively important in spite of its restricted occurrence, the low frequency of extreme storm damage, and low erosion rates. Elsewhere in the southern Gulf, some barrier shorelines show similarities to the Malpeque barrier system while others have noticeable contrasts. At Kouchibouguac Bay the significance of landward sand transport by wind and storm overwash is probably similar to that reported for Malpeque. The barrier system there (Bryant and McCann, 1973) resembles the Malpeque shoreline in coastal setting, island morphology, and long-term coastal responses. On the other hand, the low barrier islands north of Miramichi Bay in the southern Gulf have contrasts in morphology and coastal situation (Rosen, this volume). The marked differences in morphology that occur emphasize the influence that local factors can have. The important features producing the regional variability around the southern Gulf are probably the coastal wind characteristics, extreme storm effects, and sediment volumes present in the coastal zone. At Malpeque, there is a large onshore wind component, storm damage is inferred to be minor, and there is a relatively large volume of sand available along most of the barrier coast.

ACKNOWLEDGMENTS

The field program was supported by funds from National Research Council, Canada, Grant Numbers A5082 and A4227, and an award from the Science and Engineering Research Board, McMaster University, Hamilton, Ontario. Thanks are also due to Josephine Poon for typing the manuscript.

REFERENCES

Armon, J.W. (1975). Dynamics of a Barrier Island Chain, Prince Edward Island, Canada. Unpublished Ph.D. dissertation, McMaster Univ., Hamilton, Ontario, 546 p.

Armon, J.W. and McCann, S.B. (1977). Longshore Sediment Transport and a Sediment Budget for the Malpeque Barrier System, Southern Gulf of St. Lawrence. *Canadian Journal of Earth Sciences 14*, 11, 2429-39.

Armon, J.W. (in press). Morphology and Landward Sediment Transfer in a Transgressive Barrier Island System, southern Gulf of St. Lawrence, Canada. *Marine Geology.*

Bartberger, C.E. (1976). Sediment Sources and Sedimentation Rates, Chincoteague Bay, Maryland and Virginia. *J. of Sed. Petrology 46*, 2, 326-36.

Bryant, E.A. and McCann, S.B. (1973). Long and Short Term Changes in the Barrier Islands of Kouchibouguac Bay, southern Gulf of St. Lawrence. *Can. J. Earth Sci. 10*, 1582-90.

Byrne, R.J., Bullock, P., Tyler, D.G. (1975). Response Characteristics of a Tidal Inlet: a case study. *In* "Estuarine Research," Vol. 1 (L.E. Cronin, ed.). p. 201. Academic Press, New York.

Caldwell, J.M. (1966). Coastal Processes and Beach Erosion. *J. Boston Soc. Civil Engrs. 53*, 2, 142-57.

Dillon, W.P. (1970). Submergence Effects on a Rhode Island Barrier and Lagoon and Inferences on Migration of Barriers: *J. Geology 78*, 94-106.

Dolan, R. (1972). Barrier Dune System Along the Outer Banks of North Carolina: a reappraisal. *Science 176*, 286-88.

Dolan, R. (1973). Barrier Islands: Natural and Controlled. *In* "Coastal Geomorphology" (D.R. Coates, ed.), p. 263. Publications in Geomorphology, State University of New York, Binghamton.

Fisher, J.S. and Stauble, D.K. (1977). Impact of Hurricane Belle on Assateague Island Washover. *Geology 5*, 765-768.

Godfrey, P.J. and Godfrey, M.M. (1973). Comparison of Ecological and Geomorphic Interactions between Altered and Unaltered Barrier Island Systems in N.C. *In* Coastal Geomorphology(D.R. Coates,ed.),p.239.Pub. in Geomorph.,SUNY,Binghamton.

Kraft, J.C. (1971). Sedimentary Facies Pattern and Geologic
    History of a Holocene Marine Transgression. *Bull. Geol.*
    Soc. Am. 82, 2131-58.
Leatherman, S.P. (1976). Barrier Island Dynamics: Overwash
    Processes and Eolian Transport, p. 1958-1973. *In* "Proceed-
    ings, 15th Coastal Engineering Conference." American So-
    ciety of Civil Engineers, New York, 1976.
McCann, S.B. (this volume). Barrier Islands of the Southern
    Gulf of St. Lawrence.
McCormick, C.L. (1973). Probable Causes of Shoreline Recession
    and Advance on the South Shore of Eastern Long Island. *In*
    "Coastal Geomorphology"(D.R. Coates, ed.), p. 61. Publi-
    cations in Geomorphology, SUNY, Binghamton.
Moody, D.W. (1964). Coastal Morphology and Processes in Rela-
    tion to the Development of Submarine Sand Ridges off
    Bethany Beach. Unpubl. Ph.D. Diss., Johns Hopkins Univ.,
    Baltimore, 167 p.
Pierce, J.W. (1969). Sediment Budget along a Barrier Island
    Chain. *Sedimentary Geology 3*, 5-16.
Pierce, J.W. (1970). Tidal Inlets and Washover Fans. *J. of*
    Geology 8, 230-34.
Rosen, P. (this volume). Aeolian Dynamics of a Barrier Island
    System.
Schwartz, R.K. (1975). Nature and Genesis of Some Storm Wash-
    over Deposits. Coastal Engineering Research Center, Tech-
    nical Memorandum No. 61, 69 p.
Shepard, F.P. and Wanless, H.R. (1971). Our Changing Coastlines.
    McGraw-Hill Inc., New York, 577 p.
U.S. Army Coastal Engineering Research Centre (1966). Shore
    Protection, Planning, and Design. Tech. Rept. 4, 3rd Ed.,
    Washington, D.C. 580 p.

AEOLIAN DYNAMICS OF A BARRIER ISLAND SYSTEM

Peter S. Rosen

Geological Survey of Canada
Bedford Institute of Oceanography
Dartmouth, Nova Scotia
Canada

Aeolian transport of sand was monitored by using directional sand traps on the beach, foredune, overwash and spit environments of Tabusintac barrier system in northeast New Brunswick. The total volume of sand moved over the system was computed from the measured transport rates.

A large volume of sand (1720 m³) was transported in the alongshore directions from storm overwash deposits into vegetated backdune areas to produce vertical accretion. In 1977, the storm overwash deposits accounted for 8% of the subaerial volume of the system.

Net cross-island transport was offshore from overwash areas and the mid-backbeach (1556 m³) in response to prevalent offshore winds. A comparable volume (2100 m³) of sand was moved onshore from the beach to the foredune base. Approximately 220 m³ of this amount crossed accretional dune crests, but very little was transported up wave-eroded dune crests; hence only the former grew vertically. The remainder was transported alongshore in the direction of net wave-induced longshore transport and resulted in the extension of the foredune into overwash and spit areas.

Overwash deposits serve as a source of sand, and these breaches in the barrier dune also act as corridors for aeolian transport. The redistribution of storm overwash deposits by wind is important in effecting the vertical growth of a barrier island.

INTRODUCTION

The rates and total volume of sediments moved by wind were
measured on the Tabusintac barrier system, New Brunswick (Fig.1)
during summer, 1977. The purpose of the study was to relate
the variations in wind transport to barrier morphology and to
determine the role of wind transport relative to other transport agents in the barrier environment. Major sediment transfers on barrier islands are caused by nearshore, inlet, storm
overwash, and aeolian processes. The first two have major impact on the inter- and subtidal portions of the island, while
the latter two effect the subaerial growth of the system.
Aeolian and overwash are interactive processes, because overwash creates a corridor for both cross-island and alongshore
aeolian transport. Leatherman (1976a & b) identified the importance of aeolian activity in conjunction with overwash for
Assateague Island, Maryland. This field study was undertaken
to quantify aeolian movement on the subaerial environments of
a barrier island.

ISLAND SETTING

The Tabusintac barrier system is one segment of a series of
barrier spits and islands extending along the New Brunswick
coast from Miscou Island south to the Miramichi River estuary.
The tides in this region are mixed, semi-diurnal, with a mean
range of 0.8 m and a spring range of 1.2 m (Canadian Hydrographic Service, 1977). The system borders on the southern Gulf
of St. Lawrence, which is a shallow (<200 m), semi-enclosed sea
with dimensions averaging 300 by 500 km. Vanicek (1976) suggested that the Tabusintac system may be in an area of relative
sea level fall, but he noted that the data are too sparse to
make a definitive estimate. The restricted fetch of the Gulf
of St. Lawrence results in a moderately low wave energy environment during ice-free months. Wind and wave processes are
virtually stopped for a maximum of five months per year because
of snow and ice cover(Matheson 1967). The general setting of
the southern Gulf of St. Lawrence has been described by Owens
(1974) and McCann (this volume). Armon (this volume) discusses
the processes of a barrier island system in the southern Gulf
(Malpeque) with a contrasting morphology.
    The Tabusintac system is bordered on the north by Point
Barreau, a dune-capped rocky headland, and extends 14.2 km
south to Tabusintac Inlet (Fig. 1). A recently formed, ephemeral inlet (North Inlet) separates the northern section, Tabusintac Beach (6.9 km length), and the southern section, Middle

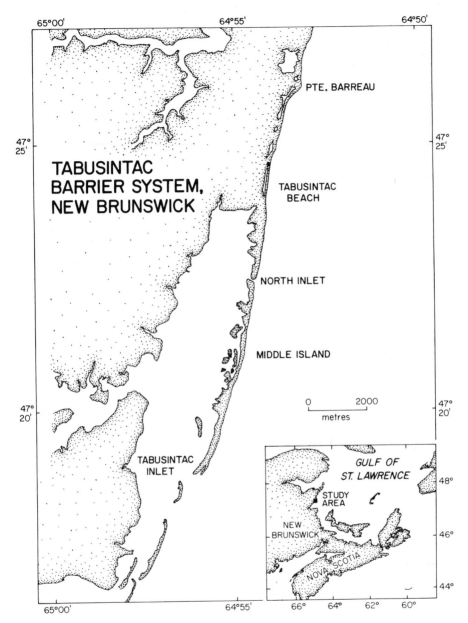

Fig. 1. Location of the Tabusintac barrier system.

Island (7.3 km length) (Fig. 2, Table 1). The Tabusintac barrier and lagoon increase in width southwards. In the wider
sections of the lagoon, numerous relict tidal delta deposits
have increased the width of the island. There is also a regional
increase in maximum dune elevation with a decrease in island
width, such that the island volume is fairly constant at any
cross section.

The supratidal morphology of this barrier island system is
simple, making it an ideal setting for an assessment of aeolian
activity. The dune system consists of a single foredune ridge
with a low-gradient, well-vegetated slope to the lagoon shore.
Thirty-five percent of the system have accretional foredunes
(a gradual seaward slope) with crest heights about 1 m above
the backbeach; 18% have erosional foredunes (dunes with vertical, wave-cut scarps) of similar height. Overwash areas are
flat, low regions formed by the flow of storm waves over the
island. Channel overwash areas (10%) are narrow, as they are
formed from a single channel of flow. Overwash plains (25%)
are wide sand flats formed from major overwash. They show evidence of numerous coalescing flow channels and may have low intermittent dunes forming on them. The remaining 12 percent is
comprised of spit deposits, similar in form to overwash plains,
but formed by the migration of inlets southwards, and concomitant island growth.

EXPERIMENTAL DESIGN

A directional aeolian sampler was used to measure total
aeolian transport (saltation and creep) up to an elevation of
45 cm. The samplers were placed in eighteen sets of four
(Fig. 3) in order to give estimates of onshore-offshore and
alongshore sediment transport rates in the following environments: overwash plain, overwash channel, erosional foredune
crest and base, accretional foredune crest and base, mid-
backbeach and spit. Samples were collected for five of the
approximately seven ice-free months. As aeolian activity was
virtually halted by snow and ice in the winter, this sampling
represented almost a full year's aeolian movement. Human and
natural interference with traps caused a shorter sampling interval in some areas but a continuous record was obtained for
all environments from May to July 1977, and for all areas except mid-backbeach and spit from May to September. In some
cases, the results from different samplers in similar environments were combined to form a continuous composite sample. Most
samplers were monitored daily in May for detailed transport,
and bi-monthly for the remainder of the study to determine total transport rates in each environment.

TABLE 1

DIMENSIONS

TABUSINTAC BARRIER SYSTEM

| | Length (m) | Width | | | | Maximum Elevation |
		Width At HHW (m)	Width At LLW (m)	Dune Width (m)	Back Beach Width (m)	Above LLW (m)
Middle Island	7,300	126.	224.	98.6	28.1	2.6
Tabusintac Beach	6,900	114.	197.	92.3	28.7	3.4
TOTAL SYSTEM	14,200	121.	209.	95.3	28.4	2.9

| | Cross-Island Volume (m^3/linear m of beach) | | | Total Volume | | |
	Above HHW	Between HHW & LLW	Above LLW	Above HHW (m^3)	Between HHW & LLW (m^3)	Above LLW (m^3)
Middle Island	76.	201.	277.	5.82×10^5	1.40×10^6	1.98×10^6
Tabusintac Beach	84.	193.	263.	5.50×10^5	1.31×10^6	1.86×10^6
TOTAL SYSTEM	80.	198.	271.	1.13×10^6	2.71×10^6	3.84×10^6

Fig. 2.   Photograph looking south across Tabusintac Beach.
Two overwash plains are in the foreground.   North Inlet and
Middle Island are in the background.

Fig. 3. Two sets of aeolian sampling units at the base and crest of a sand dune. This arrangement provides measurements of onshore-offshore and alongshore transport at each site.

The volume of sand collected in the 6.3 cm horizontal opening of the samplers was converted to the unit of cubic meters per meter of beach to express a rate of transport over the sampling period. These rates were then multiplied by the dimensions of each environment in the direction of each sampler to determine the total volume of sand transported in all directions. A net residual transport volume was computed for some environments by taking the difference of volumes transported in opposing directions.

Arrays of aluminum pins, 0.32 cm (1/8") diameter by 1 m length, were placed in each aeolian environment in order to monitor surface elevation changes. A loosely fitting washer was placed over each pin to record sequences of erosion and subsequent deposition between measurements (Rosen 1978).

The wind was monitored on the island during the sampling

period with an Aanderra weather station. Although northeast
storms form the dominant waves on this shoreline (Bryant and
McCann, 1973), the prevalent winds were strongly from the
northwest (offshore) during summer of 1977. Winds greater than
4.4 m/s (approximate initiation of aeolian movement; Bagnold
1954) were from the northwest quadrant during 38 percent of the
sampling period. Highest mean velocities (5.5 m/s) were from
the west, whereas mean velocities of 4.5 m/s were recorded from
the easterly and north-northeasterly directions (Fig. 4).

RESULTS

Transport

     In Table 2 the total transport rates measured in each en-
vironment are shown. The May-July sampling interval was unin-
terrupted for all environments, while the May-September interval was a more complete sample for all environments except
beach and spit.
     Rates of aeolian sediment transport are shown in Fig. 5.
The maximum rates were to the south, in the same direction as
net wave-induced longshore transport. The highest recorded
rate was 0.915 m^3/m of beach in overwash plains. The offshore
component of transport in overwash plains was also large
(0.540 m^3/m of beach). In the mid-backbeach significantly less
sand was moved, with highest transport being only 0.224 m^3/m
of beach offshore (east). This was measured, however, seaward
of a continuous foredune, in an area without overwash transport.
The highest transport rates on the beach were recorded at the
base of foredunes, where high alongshore transport was dominant
(up to 0.801 m^3/m of beach southwards at the base of an accretional dune). Sediment transport in vegetated backdune areas
approached zero, whereas transport on the crest of erosional
foredunes was approximately 0.005 m^3/m of beach in any
direction.
     The measured transport rates were extended over the dimen-
sions of the island (Table 3) to create a model of net onshore-
offshore aeolian transport (Table 4, Fig. 6). Overwash areas
were the sites of greatest transport even though they comprised
only 35% of the barrier system. Most sand was moved offshore
across overwash areas. The net residual sand movement from
overwash and beach areas was 1556 m^3 offshore. However, there
was a net residual movement of 459 m^3 and 1643 m^3 onshore to
the base of erosional and accretional dunes, respectively. The
volume of sand moved offshore by prevailing winds approximately
equaled the volume of sand trapped by onshore winds at the foredune base. Of this, very little sand was transported over the

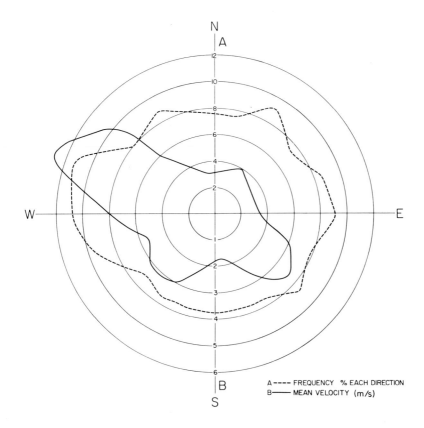

Fig. 4.   Wind regime, Tabusintac barrier, May–September 1977.

crest of erosional dunes (2.5 m³), while only a small volume
crossed accretional dunes (220 m³).   Most of the sand carried
to the foredune was transported alongshore at the dune base.
Sand accumulated here in building the foredune as well.   The
largest sediment input into backdune areas was by alongshore
movement from overwash channels (1720 m³).

TABLE 2

RATES OF AEOLIAN TRANSPORT

TABUSINTAC BARRIER SYSTEM
1977

(CUBIC METRES OF SAND PER LINEAR METRE OF BEACH)

Transport To:		Overwash Plain(x)	Overwash Channel	Erosional Dune Crest(x)	Erosional Dune Base	Accretional Dune Crest	Accretional Dune Base	Mid Backbeach	Spit
South	May-Sept.	0.915	0.314	0.005	0.593	0.104	0.801	---	(0.123)*
	May-July	0.914	0.260	0.002	0.504	0.074	0.731	0.148	0.117
West (Onshore)	May-Sept.	0.264	0.169	0.003	0.302	0.036	0.300	---	(0.134)*
	May-July	0.262	0.134	0.002	0.208	0.028	0.254	0.120	0.129
East (Onshore)	May-Sept.	0.540	0.350	0.004	0.122	0.072	0.031	---	(0.095)*
	May-July	0.504	0.250	0.003	0.074	0.057	0.026	0.224	0.088
North	May-Sept.	0.177	0.243	0.005	0.233	0.044	0.134	---	(0.166)*
	May-July	0.158	0.137	0.003	0.207	0.030	0.128	0.125	0.153

(* No August Sample)
(x Composite Sample)

RATES OF AEOLIAN TRANSPORT
TABUSINTAC BARRIER SYSTEM, NEW BRUNSWICK 1977

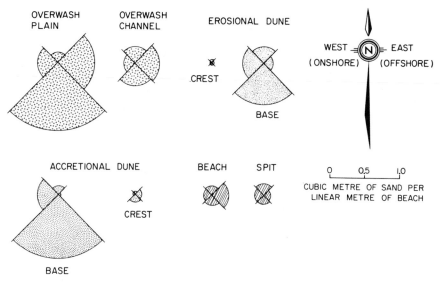

Fig. 5. Rates of aeolian transport.

AEOLIAN TRANSPORT (m³)
TABUSINTAC BARRIER SYSTEM, NEW BRUNSWICK 1977

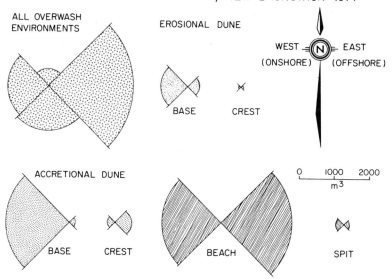

Fig. 6. Total volume of aeolian transport.

TABLE 3

GEOMORPHIC INVENTORY

TABUSINTAC BARRIER SYSTEM

	TABUSINTAC BEACH Metres	Percent	MIDDLE ISLAND Metres	Percent	TOTAL Metres	Percent
Erosional Foredune	1110.	16.0	1440.	19.7	2550.	17.9
Accretional Foredune	3650.	52.6	1485.	20.3	5100.	35.8
Channel Overwash	1230.	17.8	97.	1.3	1327.	9.3
Overwash Plain	765.	11.1	2880.	39.4	3645.	25.6
Spit	180.	2.6	1410.	19.3	1580.	11.2
TOTAL	6900.	--	7312.	--	14212.	--

TABLE 4

TOTAL AEOLIAN TRANSPORT (m³)

TABUSINTAC BEACH, NEW BRUNSWICK
1977

Transport To:	DUNE CREST			DUNE BASE			OVERWASH			SPIT[1]	BEACH[2]
	Accretional	Erosional	Total	Accretional	Erosional	Total	Channel	Plain	Total		
MIDDLE ISLAND											
South	0.83	0.04	--	3.22	2.37	--	185.82	541.48	700.30	4.49	41.63
West(Onshore)	53.46	4.32	57.78	445.54	434.88	880.38	16.4	760.32	776.72	188.94	877.0
East(offshore)	106.92	5.76	112.68	45.4	175.68	221.08	34.1	1,555.20	1,589.33	133.95	1,036.0
North	0.35	0.04	--	0.54	0.93	--	143.8	104.75	248.55	6.59	6.30
TABUSINTAC BEACH											
South	0.83	0.04	--	3.22	2.37	--	213.87	311.61	525.48	6.89	4.36
West(Onshore)	166.14	3.33	169.47	1,384.51	335.22	1,719.72	207.87	201.96	409.83	24.12	829.38
East(offshore)	332.28	4.44	336.72	141.22	135.42	276.64	430.50	413.10	843.63	17.10	979.80
North	0.35	0.04	--	0.54	0.93	--	165.51	60.28	225.79	9.30	6.50
TOTAL SYSTEM											
South	0.83	0.04	--	3.20	2.37	--	405.75	844.55	1,250.30	5.29	4.25
West(Onshore)	219.60	7.65	227.25	1,830.00	70.1	2,600.00	224.26	962.28	1,186.54	213.06	1,706.00
East(offshore)	439.24	10.02	449.40	186.66	311.1	497.76	464.45	1,968.03	2,432.75	151.05	2,016.00
North	0.35	0.04	--	0.54	0.93	--	314.0	163.67	477.37	7.14	6.44

(1) May, June, July & Sept. Only   (2) May, June, July Only

The Tabusintac barrier system is composed of 3.84 x 10^6 m^3 of sediment above the lower low water line. Of this, 2.71 x 10^6 m^3 lies below higher high water, and is the result of near-shore and inlet processes. Thirty percent of the total volume (1.13 x 10^6 m^3) laid above the higher high water line, and was deposited through overwash and aeolian activity. The total volume of sediment added to the system by overwash above the higher high water line in 1977 was estimated to be 9.03 x 10^5 m^3, or 8 percent of the subaerial island volume. While the overwash provided a source material for the building of the subaerial portion of the island, aeolian processes were significant in redistributing this sediment laterally into the backdunes effecting vertical island growth.

Elevation Changes

Repeated surveys in each environment delineated elevation changes in the topography. As illustrated in Figure 7, Profile 1, wind had the largest effect (up to 10 cm variation) on the seaward half of the island in overwash areas. The landward half of the island, with the lowest elevation, is most often inundated by high tides and saturated by ground water so that aeolian activity is inhibited (Belly 1964). A single overwash event (20 September 1977) resulted in the deposition of 20 cm of sand (maximum) on the overwash plain. The overwash deposit was found to thin toward the fan terminus, indicating the loss of transportability of the overwash surges as they move land-ward.

In the same area, wind-blown sand infilled an overwash chan-nel in less than two weeks. Profile 2 (Fig. 7) was located on the thalweg of this overwash channel. The rapid infilling of overwash channels was due to the change in surface texture. Overwash plain surfaces are typically characterized by coarse aeolian lag deposits. while the overwash channels consisted of sand. Bagnold (1954) has shown that the impact efficiency of saltating grains will decrease with a decrease in the size of bed material. Grains lose more momentum on striking sand than a coarse lag surface. The resulting decrease in total trans-port volume can be greater than 50%.

The elevation of the foreslopes of accretional dunes varied up to 25 cm in elevation, while dune-crest elevations varied a maximum of 5 cm. No elevation changes were detected landward of the dune crest. Erosional dunes showed up to 15 cm variation on the foredune slope, and no changes on the dune crest. The variations in sediment accumulation on the foreslopes in both environments were not always accretional, but responded to vari-ations in the high volume of alongshore transport. There was net accretion, however, over the sampling period in both en-vironments.

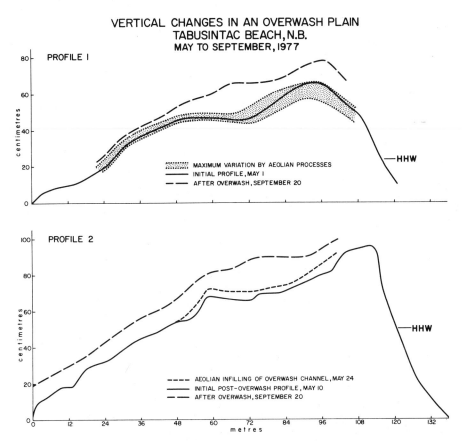

Fig. 7. Vertical changes in an overwash plain. Profile 1
was not overwashed until September 20. Profile 2 is on the
thalweg of an overwash channel formed on May 10. Note rapid
infilling of the overwash channel and dominance of overwash
deposition on the seaward half of the island. Profile 2 is
6 m south of Profile 1.

SUMMARY AND CONCLUSIONS

Maximum cross-island sediment transfer volumes were offshore due to prevailing offshore winds. However, the foredune
acts as a wind barrier, promoting accumulation of sediment from
the beach at the base of the foredune when winds are onshore.
The rate of transport is lower, but foredunes are the largest
(54%) shore-parallel feature. The residual sediment volume
transported offshore in beach and overwash areas is comparable

to the volume transported onshore to the dune base, averaged over the dimensions of the entire system. A small volume of sediment from the dune base moves onto and over the crest, but the largest volume is transferred alongshore. Sediment from the foredune slope is transported over the crest, resulting in vertical growth of accretional dunes. The wave-cut scarp on erosional dunes effectively prohibits landward sediment transport. Hence, the erosional dunes do not grow vertically until the foredune scarp has been filled in by aeolian accretion.

The high alongshore transport at the base of foredunes was significant in the building of new foredune ridges on overwashed and recently formed spit segments. Lobes of sand extended laterally from the foredune slope into unvegetated (overwash) areas. Incipient vegetation formed on these lobes over the study period. In all cases, the lobes accreted southward, in the direction of net sand transport.

Sediment transport into the backdune from the beach was minimal. The main sediment input into backdune areas was accomplished by alongshore transfers from overwash areas. Godfrey (1970) has shown the importance of overwash in the landward migration of barrier islands at Core Banks, North Carolina. This study demonstrated the significance of these washover channels as sediment sources for the maintenance and vertical growth of backdune areas. Wind-blown sand had a small effect on the sand elevation over a one year period. It was of major importance in filling and levelling overwash channels, thereby inhibiting further landward transport by overwash sediment.

Figure 8 presents a conceptual model of aeolian sediment transfers on a barrier island with prevailing offshore winds. Sand is transported alongshore from overwash areas into backdune areas; however, maximum transport volume is offshore. Net offshore transport approximately equals the volume of sand moved onshore to the base of the foredune. A small amount of sand is transported over accretional dune crests, but most is transferred alongshore, at the dune base, extending the foredune line in flat areas.

This assessment of the aeolian processes in the transport of sediment on a barrier island system demonstrates the interactive nature of aeolian and overwash processes. Storm overwash creates corridors for aeolian movement across the barrier island. While overwash processes provide the source material for building the subaerial portion of the barrier system, aeolian processes are significant for redistributing this sediment to effect vertical island growth.

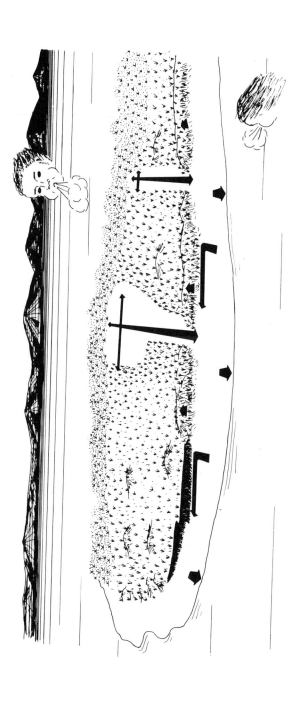

Fig. 8. A conceptual model of aeolian sediment transfer on a barrier island with prevailing offshore winds. Overwash areas input sand alongshore into backdune areas; however, maximum transport rates are offshore. Net offshore transport approximately equals the volume of sand moved onshore to the base of the foredune. A small amount of sand is transported over accretional dune crests, but most is transferred alongshore at the dune base, extending the foredune line in flat areas.

ACKNOWLEDGMENTS

E.A. Bryant and L. Carter critically read the manuscript
and made helpful suggestions for improving it. William Cooper
provided competent field assistance. Stephen Leatherman gave
useful advice on the design of the sand traps. This study was
funded by Project 770008 of the Geological Survey of Canada.

REFERENCES

Armon, J.W. (this volume). Morphology and landward sediment
    transfers in a transgressive barrier island system, Canada.
Bagnold, R.A. (1954)."The Physics of Blown Sand and Desert
    Dunes," 265 p. Chapman and Hall, London.
Belly, P.Y. (1964). Sand Movement by Wind. U.S. Army Coastal
    Engineering Research Center. Tech Memo. No. 1, 38 p.
Bryant, E.A. and McCann, S.B. (1973). Long and Short Term
    Changes in the Barrier Islands of Kouchibouguac Bay, South-
    ern Gulf of St. Lawrence. *Can. J. Earth Sci. 10*, 1582-1590.
Canadian Hydrographic Service (1977). Canadian Tide and Current
    Tables, Vol. 2, Gulf of St. Lawrence. Environment Canada,
    Ottawa.
Godfrey, P.J. (1970). Oceanic Overwash and its Ecological Impli-
    cations on the Outer Banks of North Carolina. National Park
    Service, Washington, D.C., 37 p.
Leatherman, S.P. (1976a). Barrier Island Dynamics: Overwash
    Processes and Aeolian Transport. Proc. 15th Coastal Eng.
    Conf., Honolulu, p. 1958-1974.
Leatherman, S.P. (1976b). Quantification of Overwash Processes.
    Unpubl. Ph.D. thesis. University of Virginia, 245 p.
Matheson, K.M. (1967). The Meteorological Effect on Ice in the
    Gulf of St. Lawrence. McGill Univ., Marine Science Centre,
    Montreal. Report No. 3, 110 p.
McCann, S.B. (this volume). Barrier islands of the southern
    Gulf of St. Lawrence, Canada.
Owens, E.H. (1974). Barrier Beaches and Sediment Transport in
    the Southern Gulf of St. Lawrence. Proc. 14th Coastal Eng.
    Conf., Copenhagen, p. 1177-1193.
Rosen, P.S. (1978). An Efficient, Low Cost Aeolian Sampling
    System. Current Research Part A, Geological Survey of
    Canada, Paper 78-1A, p. 531-532.
Vanicek, P. (1976). Pattern of Recent Vertical Crustal Move-
    ments in Maritime Canada. *Can. J. Earth Sci. 13*, 661-667.

A GEOBOTANICAL APPROACH TO CLASSIFICATION
OF BARRIER BEACH SYSTEMS

Paul J. Godfrey

Department of Botany
University of Massachusetts
Amherst, Massachusetts

Stephen P. Leatherman

National Park Service
Cooperative Research Unit
University of Massachusetts
Amherst, Massachusetts

Robert Zaremba

Department of Botany
University of Massachusetts
Amherst, Massachusetts

*Barrier beaches from Cape Cod to Cape Lookout have been
studied using geological and ecological techniques. The resulting data have been integrated to demonstrate the interrelationships among plants, processes, and barrier morphology. From
this analysis a general pattern for East Coast barriers undergoing recession emerges. The regional variation in vegetation
and its response to overwash has been found to be an important
criterion for classification. The vegetation of Northern barrier beaches is dominated by American beach grass (Ammophila
breviligulata) in the dune strand community, and by the decumbent form of salt meadow cordgrass (Spartina patens) in the
high marsh. Salt meadow cordgrass is killed by overwash burial
and is replaced by dune vegetation, originating from seeds and
plant fragments found in the drift lines. The stratigraphy of*

a transgressive barrier in the Northeast shows a sharp demarcation between the salt marsh and overlying washover/dune sands. In the Southeast, sea oats (Uniola paniculata) *dominates the dune grasslands, while the upright variety of* Spartina patens *is ubiquitous. The dunes along the Outer Banks develop initially as scattered clumps due to lack of well developed drift lines, irregular seed dispersal, and the clumped growth form of sea oats. There is a greater probability of overwash deposition on barrier flats and marshes in the absence of a continuous barrier dune line. The upright form of* S. patens *has the ability to grow through this overwash sediment and reestablish itself on the fan surface. An analysis of sedimentary sequences shows a deposit of clean overwash layers alternating with organic layers. These biogeological studies have shown the importance of regional variation in the vegetation in determining barrier beach topography.*

INTRODUCTION

The geomorphic development of barrier beaches (islands and spits) is greatly influenced by the type of vegetation growing on the barrier. The role of plants in the formation and stabilization of sand dunes was first demonstrated by Cowles (1899), and further studied by Olson (1958) on the Lake Michigan dunes. The ecology of dune-binding species, particularly *Ammophila*, and their role in dune formation have since been the subject of numerous studies (for example, Chapman 1949, 1976; Seneca 1969; Woodhouse and Hanes, 1966; and Ranwell 1972).

The role of grasses in stabilizing sand carried onto the back-dune area of barrier beaches by overwash was largely overlooked prior to work by Godfrey (1970). Previous investigators had noted that *Spartina patens* (salt meadow cordgrass) was a useful species to plant on open sand flats, but its role in stabilizing washovers had not been evaluated. Hosier (1973) and Travis (1976) further elucidated the adaptations of *Spartina patens* recovery from overwash burial on the Outer Banks of North Carolina. Their work on southeastern barrier islands suggests that the extensive, flat grasslands dominated by *Spartina patens* are maintained by overwash pressure during the landward migration of barrier islands.

The geomorphic role of various plant species in colonizing overwash deposits on northern barrier beaches has not been studied in detail. Leatherman (1976) showed that overwash sand is regularly transported onto the backdune areas of Assateague Island, but these deposits are not as readily colonized by *Spartina patens*, as they are in North Carolina. In the Cape Cod

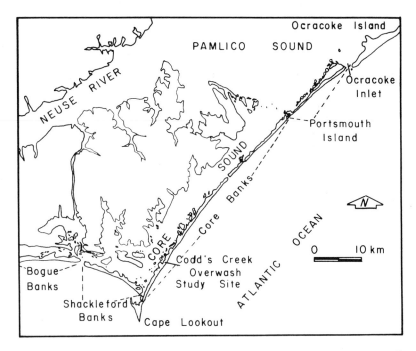

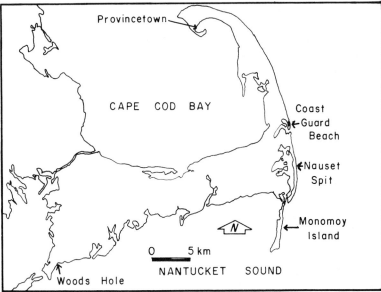

Fig. 1. Maps of Core Banks, North Carolina and Nauset Spit, Massachusetts.

area, overwash recovery seems to be linked to dune formation
rather than regrowth of grasses from beneath the washover
deposit.

These differences in ecological response by important sand-
binding species of different latitudes may influence the development of barrier beach geomorphic structures (Godfrey 1977).
We will present new data and synthesize earlier work relating
geomorphology and ecological response of vegetation on barrier
beaches subject to overwash in North Carolina and Massachusetts.
This discussion will examine the hypothesis that the structure
of barrier beaches on northern and southern coasts of eastern
North America is tied to the vegetation; we propose that such
barrier beaches can be generally classified according to their
ecological characteristics.

METHODS

Geological

A series of cores was taken along transects on Core Banks,
North Carolina and Nauset Spit, Cape Cod, Massachusetts, to
determine the stratigraphy of these barrier beaches (Fig. 1).
The coring method involved driving 7.6 cm diameter polyvinyl
chloride (PVC) pipes, up to 4 m long, into the barriers along
surveyed transect lines, using a steel pipe driver containing
18 kg of lead. When each pipe was driven as far as possible,
the above-ground portion of the pipe was filled with water and
sealed with a plumber's pipe plug. The pipe was then extracted
from the ground by means of a truck jack attached to a steel
collar or to chains wrapped around the pipe. The sample was
removed by sliding the sediment from the coring tube into a
split tube which was then opened and examined. In some cases,
the coring tube was split on a table saw using a carbide-
tipped blade. Strata were then analyzed for plant remains and
sedimentary characteristics and half of the core was preserved.

Botanical

The species composition of vegetation on washover areas was
determined by the point-intercept method, using 0.25 m^2 quadrats. Point sampling was done within each quadrat by means of
pins dropped onto the vegetation, or with a visual sighting
method that produced the same results (Grieg-Smith 1974). Both
methods provide an estimate of plant cover in each quadrat, and
with a large number of quadrats frequency can also be determined. Vegetation on Core Banks was sampled in this manner

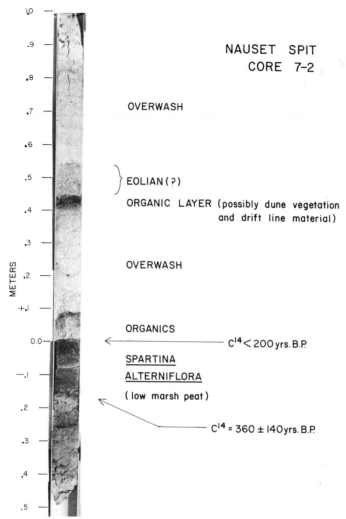

NAUSET SPIT
CORE 7-2

OVERWASH

EOLIAN(?)

ORGANIC LAYER (possibly dune vegetation
and drift line material)

OVERWASH

ORGANICS
$C^{14} < 200$ yrs. B.P.

SPARTINA
ALTERNIFLORA
(low marsh peat)

$C^{14} = 360 \pm 140$ yrs. B.P.

*Fig. 2. Photograph of the lower portion of a core from
Nauset Spit.*

along six transects, six nautical miles apart, between Barden
Inlet and Drum Inlet (Fig. 1). Thirty 0.25 m^2 quadrats were
randomly placed within 10 by 30 meter sample sites, one of
which was associated with the two to four coring locations
along each transect. Biomass samples were taken by harvesting
five 0.25 m^2 random quadrats on a 1971 overwash site and on
adjacent non-overwashed grasslands. Biomass data were treated
statistically by means of a "t" test.

Fig. 3. *Trenches through some washovers exposed cross-bedded strata, indicative of dune formation on a previous fan surface.*

Vegetation on Coast Guard Beach (Nauset Spit, Eastham), Massachusetts, was sampled by the point intercept method within a grid of quadrats placed over a washover fan and adjacent un-affected marsh. Frequency, cover and density were recorded for each 0.25 m^2 quadrat. Quadrats were placed along surveyed tran-sect lines which were tied to permanent markers for future relocation.

Elevation changes at Coast Guard Beach have been monitored since the site was overwashed during a severe winter storm in 1972. Profiles along permanent transects were taken each year to show changes due to deflation and dune building.

GEOMORPHOLOGY

The sedimentary record that documents changes in environ-ments through time can best be examined through analyses of coring data. Over 30 cores, three meters long, have been taken

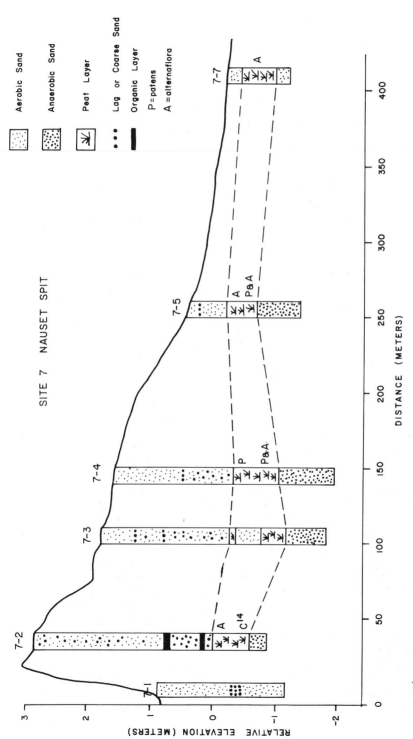

Fig. 4. Stratigraphic cross-section across Nauset Spit showing overwash sediment overlying former salt marshes.

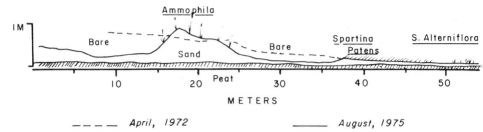

Fig. 5.   Elevation changes at Site 1 washover fan, Coast Guard Beach, due to embryonic dune development from drift line deposits.

at five transects along Nauset Spit by Leatherman (1979). Salt marsh peat was encountered near the bottom of some cores, with horizontally-stratified sand layers, characteristic of overwash deposits, found on top of the organic layers in all cases (Fig. 2). Sediment textural analysis also indicated a water-laid origin for these sandy layers.

In some vertical sections, faint cross-bedding indicative of dune strata was detected. Where the water table permitted, trenches were dug to reveal the presence of a wind-blown sedimentary section beneath some washover fans (Fig. 3). (The surface layers were always overwash sediments since the coring transects crossed washovers for topographic advantage.) A typical stratigraphic section across Nauset Spit illustrates the mechanism of barrier retreat (Fig. 4). The marsh sediments are overlain by washover deposits along the coring transect. The marsh sediments at the base of core #7-2 were dated at 360 + 125 years BP by radiocarbon methods (Geochron Labs). This date indicates a rapid rate of retreat for Coast Guard Beach, but the barrier is still a dune-dominated environment, except for some recent washover flats.

Surveys across washover features at Coast Guard Beach (Nauset Spit, Cape Cod) showed that the overwash sediment has been deflated through time (Fig. 5). However, where drift lines are present on the washover fan, sand can accumulate around and on top of the stranded algae, eelgrass, fragments and/or seeds of dune plants. The drift material thus serves as the nuclei for new dunes. Regeneration of *Ammophila* fragments and germination of seeds in the drift line facilitate dune building processes (Godfrey 1977). A portion of the overwash sand is blown to the backside of the barrier dunes by prevailing northwest winter and southwest summer winds. On most barrier beaches in the Northeast, dune environments predominate and

large open barrier flats are largely absent. Inspection of
aerial photographs show that dune and salt marsh communities
often comprise over 90% of the vegetated barrier landmass.

On Core Banks, North Carolina, the 36 cores taken by Godfrey and assistants in 1974 showed conclusively that this barrier island has been dominated by overwash during the past few
hundred years. Every core, including some over three meters
long, showed horizontal layers of sand and shells deposited by
overwash surges (Fig. 6). Layers of beach shells (including
Spisula, Donax, and *Mactra*) were found throughout each core.
Beneath and among the overwash layers, organic strata were
frequently encountered. These strata, indicative of salt marsh
peat and grassland soils, extend at least halfway across the
barrier, and are overlain by washover sediments (Figs. 7 and
8). In some cores, bay bottom sediments and shells of mud
snails (*Ilyanassa obsoleta*) were found beneath the central
portions of the barrier beach. From this stratigraphic information, it is clear that the supratidal portion of Core Banks
is an overwash structure.

Aerial photographs of Core Banks in 1939 show bay shore
marshes that have since been buried by 1.5 meters of overwash
sediment. The bay shore is now located several hundred feet
landward (Godfrey and Godfrey, 1974). Samples of wood from
in-place tree stumps found in the low tide marsh on the ocean
beach of Shackleford Banks, North Carolina (Fig. 1), were
shown to be less than 200 years old by radiocarbon dating.
Similar samples from Core Banks are believed to be less than
200 years old.

Along certain sections of the southeast United States
coast, barrier islands that are undergoing overwash retreat
tend to be dominated by extensive back-barrier flats (Fig. 7),
particularly if the islands are parallel to prevailing winds.
Dune zones tend to be rather low, scattered, and frequently
penetrated by overwash surges. While dunes can form on overwash deposits by deflation through time, the surface tends to
be stabilized quickly by vegetation recovering from overwash
burial (see next section) or by the formation of a shell lag
layer, particularly where shells are common in the overwash
deposits. Barrier islands subject to occasional overwash, or
without substantial shell concentrations in the sediment, tend
to display well-developed dune lines broken by washovers; the
breaches close by dune formation (Hosier and Cleary, 1977).
Also, barrier islands oriented across prevailing winds and/or
in regions of low sea energy will have well developed, continuous dune lines (Godfrey 1976). Such islands are rarely affected by overwash, and when they are, sand penetration is confined to the first line of dunes rather than being transported
across the entire barrier island.

Fig. 6. Cores from Core Banks illustrate the pattern of salt marsh and soil horizons interspersed with overwash deposits.

Certain barrier beaches, particularly in the mid-Atlantic region, exhibit profiles intermediate between the northern and southern overwash barriers. Assateague Island, a barrier island along the Maryland and Virginia shores, is within this transitional zone. American beach grass (*Ammophila breviligulata*) dominates the dunes, but is is noticeably less vigorous than in the Northeast. Dune environments are still an important component of the barrier system, but barrier flat grasslands are much more prevalent than in the North. *Spartina patens*, mainly a plant of the high salt marsh in the North, grows in the high marsh, on the barrier flats, and in the dunes. After overwash, drifted material is deposited along the outer edges of the washover fan, but is usually volumetrically insignificant. Recolonization of the washover fan proceeds instead by invasion from adjacent plant communities and by upward growth of buried *S. patens* (erect form) and seaside goldenrod (*Solidago sempervirens*). Several shallow cores by Leatherman (unpublished data) indicated a different sequence than was encountered in the cores from Nauset Spit. Several horizons of organic material, representing former salt marsh surfaces and grassland soils, were found to be separated by sandy overwash layers.

ECOLOGICAL RESPONSE

The back-dune barrier flats on Core Banks are dominated by *Spartina patens*, which has by far the greatest average cover (15.75%) over the areas sampled in 1974 (Fig. 8 and Table I). Associated species include *Hydrocotyle bonariensis, Erigeron pusillus, Uniola paniculata, Muhlenbergia capillaris,* and *Chloris petraea*, with average cover values between 1 and 6%. In some areas, the cover of *Spartina patens* ranged up to 66.7%, or as low as 0%, where total plant cover was very sparse, but this species exhibits a 94% frequency on the washover flats of Core Banks in the areas sampled.

The same species are generally found in the low dunes on Core Banks. Where the dunes are well developed, *Uniola paniculata* tends to dominate, but *Spartina patens* is usually present. The areas sampled are subjected to frequent overwash and thus represent a plant community adapted to this stress. Unfortunately, a substantial overwash burial of the sampled

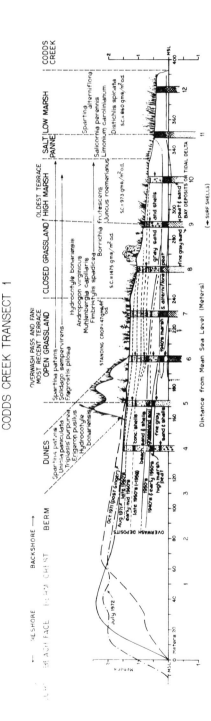

Fig. 7. Stratigraphic section at Codds Creek, Core Banks (Godfrey and Godfrey, 1974).

110

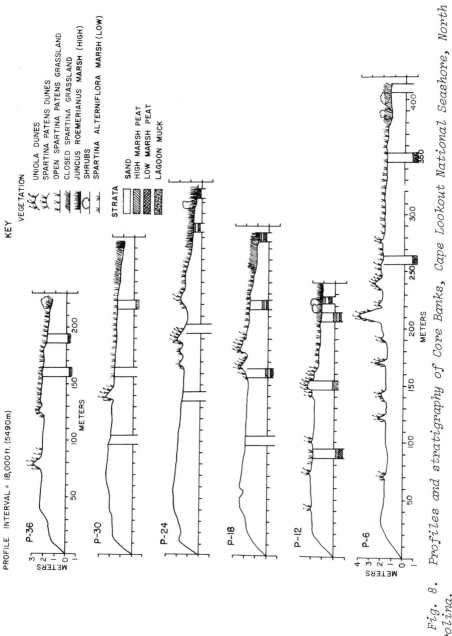

PROFILE INTERVAL = 18,000 ft. (5490m)

KEY

VEGETATION
- UNIOLA DUNES
- SPARTINA PATENS DUNES
- OPEN SPARTINA PATENS GRASSLAND
- CLOSED SPARTINA GRASSLAND
- JUNCUS ROEMERIANUS MARSH (HIGH)
- SHRUBS
- SPARTINA ALTERNIFLORA MARSH (LOW)

STRATA
- SAND
- HIGH MARSH PEAT
- LOW MARSH PEAT
- LAGOON MUCK

Fig. 8. Profiles and stratigraphy of Core Banks, Cape Lookout National Seashore, North Carolina.

111

grassland communities has not occurred since our studies began
on Core Banks, although there have been minor overwashes nearly
every year.

The ability of vegetation to recover from overwash sediment
was examined after Hurricane Ginger in 1971, when *Spartina
patens* grassland was buried by .5 meters of sediment. Portions of the grassland, not completely covered by the sand,
were sampled as a control. One year after the overwash, *Spartina patens* was growing vigorously through the deposits and had
nearly revegetated the site (Fig. 9). Excavations showed that
the regrowth was exclusively from buried vegetation. Biomass
samples taken from both the washover fan and the adjacent non-
overwashed vegetation showed no significant difference between
the two sites: 236 g/m^2 on the revegetated washover fan, and
260 g/m^2 dry weight on the non-overwashed site. The surprising
regrowth of *Spartina patens* through overwash deposits indicates
that the erect form of the species (*S. patens* var. *monogyna*)
is well adapted to overwash and plays a significant role in the
dynamics of these barrier islands (Fig. 10). Experimental
burials also showed the ability of *S. patens* to grow through
thick layers of beach sand on southeastern barrier islands
(Godfrey and Godfrey, 1976; Travis 1977).

*TABLE I. Average percent cover of overwash grassland spe-
cies from 16 plots ten meters wide and 30 meters long sampled
with random quadrats on Core Banks, North Carolina.*

Species	Average % Cover
Spartina patens *var.* monogyna	*15.75*
Hydrocotyle bonariensis	*5.80*
Erigeron pusillus	*4.30*
Uniola paniculata	*2.73*
Muhlenbergia capillaris	*2.26*
Chloris petraea	*1.14*
Ipomea sagittata	*.94*
Gallardia pulchella	*.56*
Solidago sempervirens	*.55*
Triplasis purpurea	*.55*
Andropogon virginicus	*.53*
Commelina *sp.*	*.35*
Eragrostis pilosa	*.32*
Oenothera humifusa	*.32*
Scirpus americanus	*.23*
Cynanchum *sp.*	*.21*
(ten more species, less than 0.12%)	

Fig. 9. Regrowth of Spartina patens *one year after overwash on Core Banks, North Carolina.*

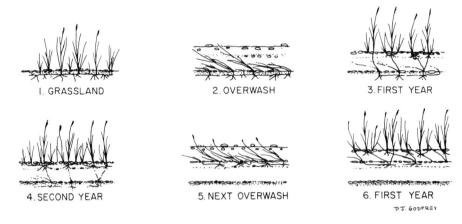

P.J. GODFREY

Fig. 10.   Diagrammatic representation of Spartina patens
(var. monogyna*) response to overwash burial and deflation on
southeastern barrier islands.*

The revegetation of overwash deposits on Northern barrier
beaches, such as Coast Guard Beach, follows a different pattern
from that recorded in North Carolina. Excavation of the 1972
washover fan on Coast Guard Beach showed that a high marsh,
dominated by a decumbent form of *Spartina patens*, had been
buried. Apparently, because the plants are weak-stemmed and
short, the *Spartina* marsh was killed by the overwash burial.
Ammophila rhizomes, culms, and seeds were washed across the
barrier during the overwash flooding and were deposited in
drift lines. Revegetation of the washover fan began from these
drift lines (Figs. 5 and 11). Fragments regenerated during the
first spring following the storm, and created a well established
dune strand community after only a few years (Fig. 12). The
dominant species colonizing the new overwash deposits was
Ammophila breviligulata (Fig. 13).

Sand, not colonized by *Ammophila* or other dune species, was
continually deflated and transported into growing dunes on the
fan, and into surviving dunes on the backside of the barrier.
This deflation created an irregular topography with low areas
subject to tidal inundation. Seedlings of *Spartina patens* were

AMMOPHILA
regenerates

3. First Year

Dune continues growth;
new sand added by later
overwashes.

6. First Year After
Overwash

Drift line with
AMMOPHILA fragments

Beach and dune sand

2. Overwash

5. Next Overwash

1. SPARTINA PATENS
High Salt Marsh

Dune grows
bare sand deflates

4. Second Year

Fig. 11. Response of vegetation to overwash and formation of dunes in the northeast.

Coast Guard Beach
Overwash Site 1
April 1972

Overwash Site 1
August 1975
(Photos- P.J.G.)

Fig. 12. Site 1 washover fan on Coast Guard Beach:
(a) April, 1972 and (b) August, 1975.

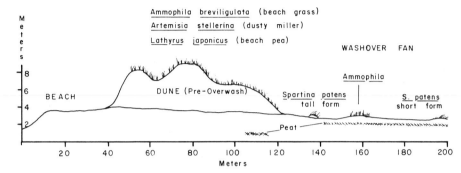

Fig. 13. Cross-section of Coast Guard Beach, Site 1 wash-over fan, showing pre- and post-1972 overwash conditions.

established at the lowest levels and also colonized the edges of the fan where the sediment was less than 10 cm thick.

A series of overwashes during 1976 added new sediment to the fan, flooded dune areas and reduced the *Ammophila* population. *S. patens* was not a significant member of the plant community that developed on the enlarged fan. The major storm of February 6-7, 1978 buried this incipient dune line completely, and the pattern of recovery is now being monitored. Other washover fans on Nauset Spit display similar patterns of recovery. Dunes develop on washovers where the fan is sufficiently elevated above daily high tides. The erect form of *Spartina*, dominant in the Southeast, was not found on these fans, although it is present in a narrow band between the high salt marsh and the base of the barrier dunes. The role of erect *S. patens* in overwash recovery in the Northeast is unknown at present, but its restricted areal distribution suggests limited importance.

Vegetation data on the Coast Guard Beach washover fan and adjacent salt marsh were collected in 1977. Trenching of the washover fan and inspection of aerial photographs showed that prior to 1972, the washover fan and adjacent high marsh were similarly vegetated. The high marsh was dominated by *Spartina patens* with a cover of 72%; frequency of 89% (Table II). However, vegetation on the washover fan, which was sparse compared to the marsh, was dominated by *Ammophila breviligulata*: cover 0.58% and frequency 11.2% after the 1972 overwash. *Spartina patens* was present in the lower portions of the fan, and was derived from seedlings; its cover was 0.41% and its frequency, 5.3%. Other species found on the fan were typical dune and beach plants such as *Lathyrus japonicus, Artemisia caudata, Artemisia stelleriana, Salsola kali, and Cakile edentula.*

TABLE II. Average percent cover of species on overwashed and non-overwashed adjacent sites on Coast Guard Beach, Mass.

| | Percent Cover | |
	Non-overwashed high marsh	Overwash fan on high marsh
Bare sand	*10.69*	*82.03*
Organic drift	*2.28*	*16.86*
Spartina patens	*72.26*	*0.41*
Spartina alterniflora	*18.57*	*0.00*
Salicornia virginica	*2.53*	*0.00*
Puccinellia maritima	*1.59*	*0.00*
Limonium nashii	*0.39*	*0.00*
Agropyron repens	*0.09*	*0.01*
Distichlis spicata	*0.07*	*0.00*
Suaeda maritima	*0.06*	*0.00*
Ammophila breviligulata	*0.00*	*0.58*
Cakile edentula	*0.00*	*0.04*
Artemisia stelleriana	*0.00*	*0.03*
Artemisia caudata	*0.00*	*0.01*
Lathyrus japonicus	*0.00*	*0.01*
Salsola kali	*0.00*	*0.01*

DISCUSSION AND CONCLUSIONS

The geomorphology of transgressive barrier beaches in the Northeast differs significantly from that in the Southeast (Outer Banks). In the Northeast, dune ridges and overall dune-type topography dominate the barriers, whereas in the Southeast, landward migrating islands in their natural state have extensive washover flats, and open, scattered dune fields.

While major differences can be seen in general topography, the basic processes of sand transport along these coasts are similar. Transgressive barriers are characterized by overwash deposits overlying salt marsh sediments of approximately the same age. In both the Northeast and Southeast, overwash creates the basal surface on which dunes form. However, the tendency for continual dune development on washover fans seems to be much greater in the Northeast than in the Southeast, where the overwash covers previous grassland communities, or shell lag layers develop. The frequency of overwash depends on the tidal range and storm exposure. The small tidal range along the Outer Banks and the high storm frequency result in numerous overwashes. In addition, poor dune development in east- or southeast-facing barriers favors overwash. The greater tidal

Fig. 14. (a) Dune development on Nauset Beach, Massachusetts, showing dense and continuous cover of Ammophila breviligulata with lateral growth into open sand. (b) Uniola tends to form dune clumps rather than distinct barrier dune lines on Core Banks, North Carolina.

range in the Northeast results in a lower probability of overwash for a storm of a given size. Only during large, slow-
moving storms, such as the February 6-7, 1978 northeaster,
does major overwashing occur. This storm created extensive
washovers on Coast Guard Beach, providing abundant sediment
for new dune formation.

A typical profile of a Northeastern barrier beach shows
high, often scarped, dune ridges close to the beach, backed by
a second zone of dunes or dune ridges (Fig. 13). These dunes
grade into an intertidal salt marsh. The supratidal zone dominated by dunes tends to be narrow on retreating barriers. On
the other hand, the typical profile of Southeastern barriers,
such as the Outer Banks, is wide and low, with extensive back-
barrier flats, low, open dune zones, and wide beaches (Figs.
7 and 8). Such profiles are, of course, modified where man-
made dunes have been created.

The vegetation of barrier beaches changes significantly
from the Northeast to the Southeast. On northern barriers,
Ammophila breviligulata is the dominant dune stabilizer, producing extensive dense grasslands in regions above the intertidal zone. *Ammophila* grows best where new sand is continuously
added to the dune and where flooding is infrequent. *Ammophila*
can grow vertically with rapidly accreting dunes, as well as
laterally into open sand flats. Rapid growth of *Ammophila* (up
to 2 mm/day at Cape Cod; Brodhead and Godfrey, 1977) results
in substantial, densely vegetated dunes (Figure 14a). *Ammophila*
grows best in New England, with less vigor near its natural
southern limit in North Carolina. On the Outer Banks, *Ammophila*
is subject to heat stress, disease, and insect attacks, and
must be heavily fertilized to produce the typical dense stand
characteristic in the northern part of its range.

In the South, where *Ammophila* grows poorly, *Uniola paniculata* dominates the dunes. *Uniola* is better adapted to heat
stress and the environmental conditions of southern barrier
beaches. It does not grow laterally as rapidly as *Ammophila*,
although it grows upward at least as well. *Uniola* also does
not tend to create as dense a stand as *Ammophila*; *Uniola* dunes
subject to frequent overwash tend to form open dune fields,
such as those on Core Banks (Fig. 14b). Washovers, stabilized
by the rapid recovery of grassland vegetation, do not act as a
continuing sand source for dune building, particularly if prevailing winds blow across the washover fans toward the ocean.
Where barrier beaches are oriented across prevailing winds
with open sand beaches, well developed dune systems form even
with the somewhat limited growth habit of *Uniola*.

Changes in local distribution of *Spartina patens* and its
varieties along the East Coast play a significant role in the
stabilization of overwash deposits and thus affect the geomorphology of barrier beaches.

SOUND   OCEAN

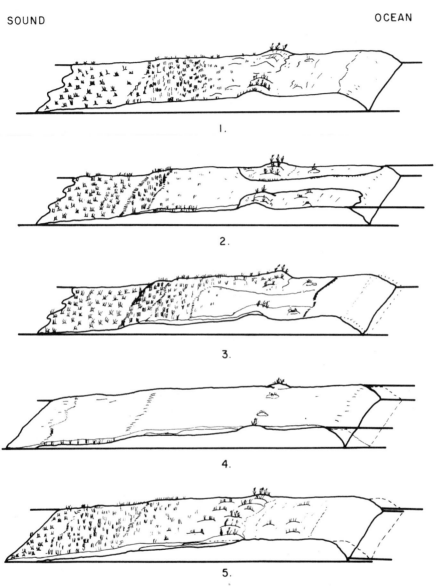

Fig. 15. Model of overwash response and retreat in the southeast (Godfrey and Godfrey, 1970, 1976).

The southern form of *Spartina patens* (var. *monogyna*), a tall, robust, and rapidly growing plant, can tolerate sea water flooding and overwash burial. It dominates the back-barrier grasslands of the Outer Banks and can also form dunes. This

species is probably the most important plant on the rapidly
retreating Southeastern islands. Wherever healthy stands of
S. patens var. *monogyna* are present, overwash deposits are rapidly colonized by the upward growth of buried plants. Burial
of erect *S. patens* stimulates upward growth, and leads to a
dense stand within a few years. As the grassland grows, deflation from overwash deposits is reduced and eventually prevented. The formation of shell lag layers on the surface of washovers also prohibits further deflations leading to broader back-
barrier flats typical of southeastern washover barriers. Figure
15 still appears to be a valid model for the response of vegetation on barriers such as Core Banks. The model by Hosier and
Cleary (1977), based on their study at Masonboro Island, North
Carolina, corroborates this work.

Toward the North, the predominance of the tall form of
Spartina patens declines, but this form does extend along the
Virginia coast to Delaware, into New Jersey and New England.
The form most commonly found in the North is the decumbent
type: small, weak-stemmed, and incapable of surviving substantial sand burial. The short and tall forms grow side by side
on parts of the New York coast and in Massachusetts. On Coast
Guard Beach, and elsewhere in New England, the tall form (var.
monogyna) occupies a few localized bands between the high marsh
and dune zone, often occurring on low dunes at the edge of the
marsh. The roles of the two forms in sand stabilization of
northern barriers are presently being investigated by John
Brodhead, Mark Benedict, and the authors, all of the University
of Massachusetts. Data to date indicate that the northern decumbent form of *S. patens* is not overwash-adapted, and dies
when covered by substantial overwash deposits (over 30 cm).
Dune strand vegetation is associated with overwash recovery in
the North, rather than salt marsh vegetation. The decline in
importance of *Spartina patens* as a grassland species parallels
the increase in importance of *Ammophila* as the primary dune-
building species. In the mid-Atlantic region both plants grow
well. This region also shows a change in geomorphology from
the dune-dominated barriers of the North to the barrier flats
of transgressive islands in the Southeast. We suggest that this
change in geomorphology is partially a result of the differences
between the stabilizing capabilities of *Ammophila* and *S. Patens*.

The model which has been proposed for the stabilization of
Northeastern barrier beaches, following overwash, is shown in
Figure 16. The diagrams show the progressive erosion of a substantial dune ridge as the shoreline retreats. Eventually the
dune line breaks, and overwash sand is carried onto the back-
barrier, burying the high marsh behind. Drift material carried
in storm surges is deposited with overwash sediment. During
early spring, fragments and seeds of dune strand species,

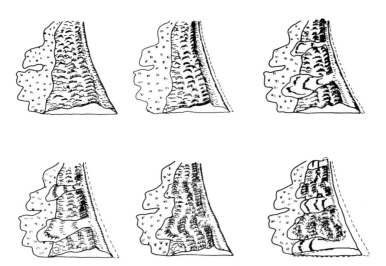

Fig. 16. Model of overwash response and retreat in the Northeast.

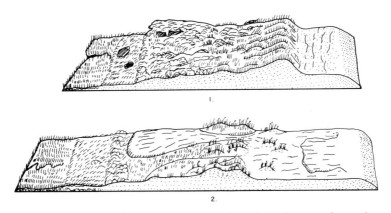

Fig. 17. General types of barrier beaches undergoing landward retreat: 1. Northeast, and 2. Southeast.

primarily *Ammophila*, begin to grow in scattered locations. Prevailing northwest and southwest winds deflate open sand areas on the washover fan. Since the washover deposit lacks large quantities of shell or gravel, no substantial lag deposits form and deflation continues. Sand from the fan is blown into newly-forming dunes, and is trapped by sand-binding dune plants. Since revegetation does not occur from below, as it does in the Southeast, vegetative cover is discontinuous for a long time, with dune patterns reflecting the initial drift line deposition.

Stabilization of washovers appears to follow a cycle of events related to oceanic conditions and plant response. An occasional massive overwash in the Northeast will create an extensive surface on which new dunes can form. Rapid growth of *Ammophila* leads to the development of substantial, stable dune lines, that can resist most overwashes. Time between major overwash events appears to be considerably longer in the North than in the South, at least in the recent past. In the Southeast, rapid revegetation of washover fans by *Spartina patens*, and little deflation of overwash deposits, results in more open, flat topography than that found in the North. Flat topography leads to an increased likelihood of overwashing. On the unmanipulated Outer Banks, the overwash/stabilization cycle seems therefore to be more rapid than that in the North.

These differences suggest that barrier beaches can be classified according to certain ecological characteristics. East Coast transgressive barriers can be divided broadly into two types, related to vegetation as well as topography (Fig. 17). Type 1 is common for Northeast barriers, with dune-dominated topography stabilized by *Ammophila breviligulata*. Dunes have formed on overwash deposits. Type 2 represents the typical barrier of the Southeast Coast (Outer Banks): wide, with a low, open dune zone, and extensive barrier flats dominated by *Spartina patens* grassland. This type of barrier tends to be dominated by overwash, with sporadic dune growth. Both types of barriers shown in the figure face east and experience high energy seas from frequent storms.

The difference in the ecological response of major sand-binding plants suggests that management approaches should be developed according to the conditions in a given region. On Northern barriers, planting *Ammophila*, along with sand fencing in open sand areas, would encourage dune formation and aid the natural overwash recovery process. In the Southeast, a better policy would be to encourage the development of overwash-adapted grasslands dominated by *Spartina*. Dune regions should remain open so that the barrier can be overwashed whenever possible. The stabilizing grasslands can survive sea water flooding, and provide a surface on which overwash sand can be deposited, building the back-barrier regions as sea level

rises. Such a policy, of course, will create other problems
on heavily-developed barrier islands.

This study shows several important relationships between
plants and geological processes, particularly related to
barrier beaches. We feel that it is important for geologists
and ecologists to recognize that different ecological mechanisms exist in response to similar geological processes along
the East Coast. A complete understanding of barrier beach
dynamics requires that the ecological responses of important
sand-binding species be considered along with geological processes. We hope that further geological treatments of barrier
beaches will consider the plants involved and test this
hypothesis.

ACKNOWLEDGMENTS

We extend our thanks to the Office of the Chief Scientist,
National Park Service, Washington, D.C.; the North Atlantic
and Southeast Regional Offices of the National Park Service;
and the Superintendents and Staff of Cape Lookout, Cape
Hatteras, Assateague Island, and Cape Cod National Seashores.
We thank Melinda Godfrey for her editorial review of this
manuscript.

REFERENCES

Brodhead, J. and Godfrey, P.J. (1977). Effects of Off-Road
    Vehicles on Plants of Dunes at Race Point, Cape Cod Na-
    tional Seashore. NPS-CRU Report No. 32.
Chapman, V.J. (1949). The stabilization of sand dunes by vege-
    tation. Proc. Conf. Biol. and Civil Eng., p. 142-157.
    Inst. Civil Eng., London.
Chapman, V.J. (1976, 1964). Coastal Vegetation, 2d ed., Mac-
    Millan Co., New York.
Cowles, H.C. (1899). The ecological relations of the vegeta-
    tion on the sand dunes of Lake Michigan. Bot. Gaz. 27, p. 95-
    117, 167-202, 281-308, 361-391.
Godfrey, P.J. (1970). Oceanic overwash and its ecological im-
    plications on the Outer Banks of North Carolina. Ann. Rpt.
    1969. Office of Natural Science Studies, National Park
    Service, U.S. Dept. of Interior, p. 1-37.
Godfrey, P.J. (1976b). Comparative ecology of East Coast bar-
    rier islands. In Technical Proc. of the 1976 Barrier Island
    Workshop. The Conservation Foundation, Washington, D.C.,
    p. 5-34.

Godfrey, P.J. (1977). Climate, plant response, and development
    of dunes on barrier beaches along the U.S. East Coast.
    Intern. J. Biometeor. 21, p. 203-215.
Godfrey, P.J. and Godfrey, M.M. (1974b). The role of overwash
    and inlet dynamics in the formation of salt marshes on
    North Carolina barrier islands. *In* "Ecology of Halophytes,"
    Academic Press, N.Y., p. 407-427.
Grieg-Smith, P. (1964). Quantitative Plant Ecology, 2d ed.
    Butterworth, Washington, D.C.
Hosier, P.E. (1973). The effects of oceanic overwash on the
    vegetation of Core and Shackleford Banks, North Carolina.
    Ph.D. thesis, Duke University.
Hosier, P.E. and Cleary, W.J. (1977). Cyclic geomorphic pat-
    terns of washover on a barrier island in southeast North
    Carolina. *Environmental Geology 2*, p. 23-32.
Leatherman, S.P. (1976). Barrier Island Dynamics: Overwash
    Processes and Eolian Transport. Proc. of the 15th Coastal
    Engineering Conference, Honolulu, p. 1958-1974.
Leatherman, S.P. (1977). Assateague Island: A Case Study of
    Barrier Island Dynamics. Proc. of the First Conference on
    Scientific Research in the National Parks, New Orleans,
    17 p.
Leatherman, S.P. (1979). Landward Migration of Nauset Spit:
    Relative Rates and Mechanisms. NPS-CRU-UMass Report
    (in press).
Olson, J.S. (1958). Lake Michigan dune development II: Plants
    as agents and tools in geomorphology. *J. Geol. 66*, p. 345-
    351.
Ranwell, D.C. (1972). Ecology of salt marshes and sand dunes.
    Chapman and Hall, London, 258 p.
Seneca, E.D. (1969). Germination response to temperature and
    salinity of four dune grasses from the Outer Banks of
    North Carolina. *Ecol. 50*, p. 45-52.
Travis, R.W. (1976). Interaction of plant communities and
    oceanic overwash on the manipulated barrier islands of
    Cape Hatteras National Seashore, N.C. Proc. of Conf. on
    Scientific Research in the National Parks, New Orleans,
    17 p.
Woodhouse, W.W. and Hanes, R.E. (1966). Dune stabilization with
    vegetation on the Outer Banks of North Carolina. Soils
    Info. Ser. 8, North Carolina State University, Raleigh,
    50 p.
Zaremba, Robert (1979). The Ecological Effects of Off-Road
    Vehicles on the Beach/Backshore Driftline Zone in Cape
    Cod National Seashore, Mass, NPS-CRU Rpt. No. 29 (in
    prep.).

WASHOVER AND TIDAL SEDIMENTATION RATES AS
ENVIRONMENTAL FACTORS IN DEVELOPMENT OF A
TRANSGRESSIVE   BARRIER SHORELINE

John J. Fisher
Elizabeth J. Simpson

Department of Geology
University of Rhode Island
Kingston, Rhode Island

*Washover fans and tidal deltas are significant reservoirs
of sediment in barrier island systems. These back-barrier and
lagoon deposits also represent important sediment sinks in the
overall littoral sediment budget of a shoreline. Along the
40 km long Rhode Island south shore barrier beaches, extensive
washover fans, resulting from hurricanes, and several inlet deposits are prevalent. A long-term quantitative analysis of the
subtidal and supratidal sedimentation of these units from 1939-
1975 was conducted using photogrammetric techniques. Back-
barrier sedimentation features were first identified in the
field to develop "ground-truth" keys to identify the features
on the aerial photography. Sediment units were measured on each
set of aerial photographs by point-counting of grids. These
photographic measurements were scaled by using ground survey
data.*

*During the 36-year study period, the total areal sedimenta-
tion change of the supratidal washover deposits was +522,790 m^2.
The subtidal washover deposits amounted to +267,950 m^2, for a
total washover sedimentation accumulation of 790,740 m^2. For
the same time period, the total area change of supratidal flood
tidal deltas at the inlets was +188,240 m^2, while that of the
subtidal tidal deposits was +862,320 m^2 for a total tidal delta
deposition of +1,050,560 m^2. This analysis indicates that during this 36-year period, tidal delta sedimentation is 1 1/3
times more effective than washover in the landward transportation, deposition and storage of sediment.*

*This barrier coast, in addition to transgressing, is eroding
at the rate of 0.7 m/yr. An inverse relationship exists between*

overwash occurrence and barrier island width, which can be fur-
ther related to beach erosion. Measured beach erosion over
the 36-year period is directly related to the rate of overwash
occurrences at different point along the coast.

INTRODUCTION

     Accumulated evidence tends to indicate that many present-
day barrier island coastlines have developed under a eustatic
rising sea level and thus on a shoreline of submergence. Development of this shoreline then leads to transgressive deposits as the subsurface record. The early Davisian model suggested that a barrier island would migrate landward by tidal inlet
deposits, dune migration and overwash deposits. It now appears
that the most significant of these for present-day transgressing barriers may be that of overwash. Thus under a rising sea
level, extensive thick overwash deposits might be common in the
stratigraphic section. The purpose of this study was therefore
to determine the rate by which these deposits develop on a
present-day barrier beach in both the subtidal and supratidal
coastal environments. In order to make the study more statistically valid, the entire Rhode Island barrier beach shoreline
was studied rather than studying a portion of the shoreline and
extrapolating "average" values to the remainder of the coast.
     The Rhode Island coast (Fig. 1) is considered very suitable
for this isolating of factors since it is of moderate length
(40 km) and does not have any input of sediments by rivers,
either to the coast or the lagoons. While the longshore current regime is moderate (from west to east), there does not
appear to be any prominent longshore drift sinks.
     Two depositional processes active along the barrier beaches
of the south shore of Rhode Island are washover and tidal delta
sedimentation on the back-barrier and in the lagoons and ponds.
The observed general trend of washover and tidal delta sedimentation on these barrier beaches is to deposit sediment on the
back-barrier or in the adjacent lagoons. On a generally erosional shoreline such as the Rhode Island south shore, the result of beach erosion and washover deposition is the landward
migration of the barrier beach system. If sufficient sediment
is supplied to the barrier beach system by washover, tidal delta
and dune sedimentation to balance offshore and downdrift beach
losses, then the barrier beach coastline will maintain itself.

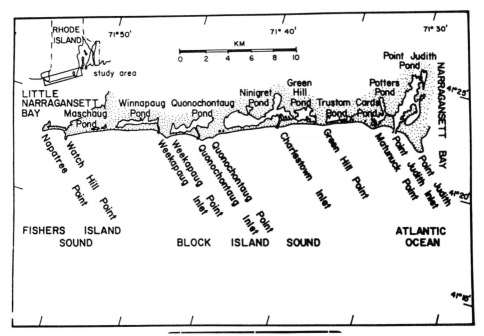

Fig. 1. Barrier beach shoreline of southern Rhode Island developed on the outwash plain from a Wisconsian recessional moraine. Headlands, locally called "points," composed of till and till/bedrock, separate the barrier beaches developed across the former outwash channels. Four stabilized inlets are presently open in these barriers.

REMOTE SENSING PHOTOGRAMMETRY ANALYSIS

To determine the rate of overwash deposition, a range of aerial photographs covering the entire Rhode Island coast were mapped photogrammetrically for the period covered by aerial photography from 1939 (oldest) to 1975 (most recent). The 1938 coverage was several months after the 1938 hurricane, considered the most intense in recorded history. These aerial photographs were interpreted both in the laboratory and in the field to determine photogrammetrically the changes in areas of the back-barrier and lagoon deposits, and to distinguish the changes as washover or tidal delta sedimentation. Changes along the beach, measured as the high water line, indicate the possible amount of sediment that is available to be transported landward by overwash and tidal inlet processes. Backbarrier sedimentation resulting from landward eolian transport of sediment was not distinguished from overwash transport of sediment on the basis

of aerial interpretation. Eolian transport was considered less
significant than overwash sedimentation on the basis of Bar-
theger's study (1976) of sedimentation in Chincoteague Bay
from Assateague Island, Virginia - Maryland. In addition,
Leatherman (1977), from short-term field studies, reported a
small net loss from washover fans due to eolian transport by
the prevailing offshore winds at Assateague. This study was
designed to determine the regional sediment budget so that only
the total amount of sedimentation over the long term was considered.

Amounts of areal changes of subtidal and supratidal overwash and inlet tidal deposits and high water shorelines were
measured using a standard grid point-count technique on stable
mylar overlays, rectified as to scale and ground controls with
a Bausch and Lomb Zoom Transfer Scope. Linear measurements
were made with a direct reading micro-rule to a precision of
.001 inch.

The Bausch and Lomb Zoom Transfer Scope (Fig. 2) is an optical, anamorphic copy system that allows ratio comparisons of
1:14 and was initially designed to permit overlays of larger
scaled satellite imagery on small scaled topographic maps.
Early photogrammetric shoreline change studies (e.g., Stafford
1972, 1971) required rectified enlargements of each aerial photograph. Direct measurements or overlays on unrectified aerial
photographs may produce errors due to scale differences (sometimes as great as 5 percent), camera tilts, radial displacement,
print paper shrinkage, and relief displacement. A photographi-
cally rectified print can reduce these errors, but is is ex-
pensive. The anamorphic feature of the Zoom Transfer Scope
allows on-line displacement of the photographic image in both
z and y axis and thus can optically rectify each overlay
image to stable ground control points. In addition, the continuous enlargement system allows each aerial photograph to be optically enlarged to exactly the scale of ground control survey
points. Several ground control survey points were used for
each aerial photograph. Comparison of aerial and ground control measurements indicated an accuracy of 3 m.

Quantitative analysis requires that the order of accuracy
of photogrammetric measurements be related to the true field
values. To reduce micro-rule errors, scale and cartographic
variability, and operator variability, a ground-truth survey of
linear distances and areas was made. Linear distances were
measured in the field at several localities at the same elevation. These linear ground measurements were used to calculate
the ground areas of rectangular fields and buildings, to be
used as ground-truth values for both linear distances and areas
that were then measured on a 1972 photograph (073-72 series,
with a nominal scale of 1:12,000). The quantitative amounts of

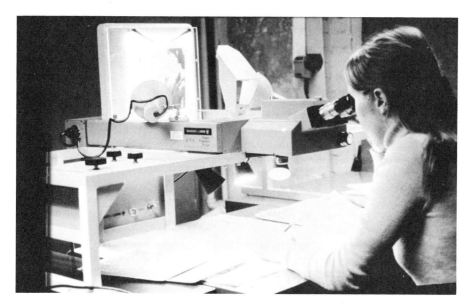

Fig. 2. Areal changes of backbarrier sedimentation were photogrammetrically measured using aerial photographic overlays from 1939 to 1975 on a Zoom Transfer Scope. The Bausch and Lomb Zoom Transfer Scope, an optical enlarging/reducing copy system, possesses anamorphic capabilities that can correct optically the usual unrectified aerial photographs.

error or variance resulting from the linear and areal measurements of the objects of known areas (determined from the field measurements) average 2.1%.

RHODE ISLAND BARRIER BEACHES

The south shore of Rhode Island (Fig. 1) is a submergent coast developed on a glacial outwash plain with the former glacial outwash channels now enclosed by barrier beaches built by wave action. These barrier beaches are separated from each other by low headlands of till or till-covered bedrock. The four inlets in these barrier beaches are all stabilized by jetties and are locally called "breachways." These barrier beaches genetically might be classifed as "baymouth bars" using the Johnsonian descriptive classification. The term "barrier beach" was more recently defined by Shepard (1952) as a single elongate sand ridge above high tide and parallel to the

coast and lagoon. It is differentiated from a barrier island
("multiple sand ridges") and a barrier spit ("attached at one
end to mainland"). Thus these Rhode Island barriers might further descriptively be referred to as "barrier spits," since
each is attached at its base to the glacial headlands.

Dillon (1970), in a subsurface study of one of these lagoons, discussed the effect that a rising sea level had in sub-
merging this Rhode Island barrier beach and causing a landward
migration--a "roll-over" of the barriers across their lagoons.
Although Dillon discussed this "roll-over" effect for the Rhode
Island barriers specifically, it has been applied to other barrier islands. This migrational process is accomplished by
sedimentation processes that transport sediment from the front
to the back of the barrier. Flood tidal delta deposition at
the inlets is one of the processes in action along this coast
that moves this sediment landward. The tidal delta at the
Charlestown Breachway (Fig. 3) is a final depository for incom-
ing sediments since the inlet flood current velocities are high
compared to any currents in the lagoon.

Washovers along the Rhode Island barrier shore also move
sediment landward (Fig. 4). During the 1938 hurricane, overwash surges eroded the barrier dunes, depositing "great scallops
of sand which extend out over the marsh as much as 750 feet from
the eroded foredune" (Nichols and Marston, 1939). An increase
in sand size behind the barrier after a hurricane was considered
by Dillon (1970) as evidence for overwash. Eolian action on
these overwash fans and channels may move sediment landward or
seaward depending on the direction and speed of coastal winds.

Prevailing winds on the Rhode Island coast are from the
southwest, west, and northwest. Storm winds approach from
every direction except the southwest, with the northwest favored for duration and velocity. Winds from the southeast are
rare, but their infrequency is more than compensated for by
their severity, with a maximum damaging effect for each occurrence. Swells are predominantly from the east but low swells
from the northeast, southeast, south and southwest are also
common. Medium (2.0-4.0 m) and high (greater than 4.0) swells
are predominantly from the east (U.S. Army Beach Erosion Board,
1949).

PREVIOUS STUDIES

Washovers

The actual process of overwash has been described as a uni-
directional, discontinuous flow or pulse of sediment-charge
water which occurs in response to the storm wave runup and

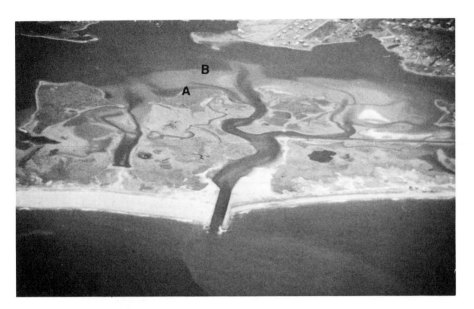

Fig. 3. Flood tidal delta of Charlestown Inlet, one of four similar inlets along the Rhode Island barrier beach coast. Supratidal (A) and subtidal (B) flood tidal delta backbarrier shorelines can be identified and mapped photogrammetrically on vertical aerial photographs to determine quantitatively sequential shoreline changes.

storm surge overtopping the barrier (Schwartz 1975; Fisher, Leatherman, and Perry, 1974). Overwash is generally accepted as
occurring as a result of the combined effects of a storm surge,
storm waves, and normal or unusually high tides.

The plane geometry of a washover deposit is controlled by
the degree of foredune development and the backbarrier topography. Overwash localized at isolated low points in the foredune
(such as a blowout, former overwash channel, or beach buggy
access road) generally produces distinct washover fans . If a
fairly continuous section of the foredune is lower than the combined level of the storm surge, wave heights, and tide level,
then overwash will occur over this shoreline length, producing
a coalescing or sheetlike deposit of sediment on the backbarrier
(which has been termed a washover ramp). Backbarrier topography
can affect the washover geometry by serving to disperse or contain overwash surge and sediment deposition (Schwartz 1975). No
photogeologic studies are known which have attempted to quantify
widespread backbarrier shoreline changes in the manner of this

Fig. 4. Washovers along the western portion of the Rhode Island shoreline developed across the barrier beaches into Maschaug Pond. Washover channels extend from the beach, through the foredune across to the lagoon. The backbarrier supratidal shoreline of these washovers in aerial photographs have a characteristic bulbous fan shape.

study. Stirewalt and Ingram (1974) measured shoreline changes over a thirty-year period along the mainland salt marsh environment in Pamlico Sound, North Carolina, but only at isolated localities.

Washover deposition appears to have an important long-term effect on barrier beach evolution in maintaining the existence of these landforms. Overwash is recognized as a process that rejuvenates the backbarrier marsh by creating new marsh fringe (Godfrey 1976; Godfrey and Godfrey, 1973; Scott, Hoover and McGowen, 1969), although continuous, excessive washover deposition can result in the destruction of the marshes by exceeding their capacity to recover and benefit from the input of new substrate (Godfrey 1976). Washover deposition on the backbarrier also results in the vertical accretion of sediment in addition to the lateral accretion which occurs at the marsh or lagoon fringe (Nordquist 1972; Godfrey and Godfrey, 1973; Godfrey 1976; Leatherman 1976).

The terms "overwash" and "washover" have occasionally been used interchangeably in the literature, but a distinction drawn by Schwartz (1975) between the terms is followed here. Schwartz defined "overwash" as the mass of water that overtops the barrier island, as well as the process of overtopping; and "washover" as the sediment deposit or geomorphic features produced by the process of overwash. "Backbarrier" as used in this study refers to the area between the foredune ridge and the subtidal boundary of the lagoon or bay shoreline.

Tidal Deltas

Accumulations of sediment develop at both the oceanward and lagoonward ends of an inlet in response to the tidal flow through the inlet, wave climate, and supply of sediment from longshore drift, fluvial discharge, and the offshore. Hayes and Kana (1976) identified three principal sand units associated with inlets: flood tidal deltas, ebb tidal deltas, and recurved spit-inlet fill sediments associated with inlet migration. The degree of development of flood and ebb tidal deltas appears to be a function of the amount of sediment supplied to the inlet area, and of the interaction of waves, longshore drift, tidal currents, and fluvial discharge.

The relationship of geomorphic and sedimentologic aspects of tidal deltas has been studied by Lucke (1934a, 1934b), Fisher (1962), Hoover (1969) and Caldwell (1972). The interrelationship of marshed, washover, and tidal delta sedimentation in the evolution and development of the barrier beach system has been studied by Pierce (1970), Godfrey and Godfrey (1973, 1974, 1975), and Godfrey (1976). The evolutionary development of tidal deltas has been considered by Morton and Donaldson (1973), who

developed and interpreted Lucke's (1934a, 1934b) three stages
of tidal delta development, and by DeAlteris (1976), who considered lagoon infilling as a factor of inlet and lagoon hydraulics.

Noting that tidal range appears to have the broadest effect
in determining large-scale differences in estuary or lagoon sand
accumulations of all the process variables (tidal range, tidal
currents, wave conditions, and storm action), Hayes and Kana
(1976) consider three coastal models: microtidal (tidal range
less than 1 m), mesotidal (tidal range 1-4 m), and macrotidal
(tidal range greater than 4 m). On microtidal coasts the effects of waves generally dominate over those of the tide. For
this reason ebb tidal deltas are largely absent or poorly developed. Tidal currents generated through the inlet are generally capable of transporting only fine sediment or relatively
insignificant amount of sediments, which tend to be redistri-
buted by the waves before it can accumulate. Flood tidal deltas
on a microtidal coast are also generally poorly developed, although they tend to be better developed than their associated
ebb deltas. On mesotidal coasts both the ebb and flood tidal
deltas tend to be well developed and have characteristic mor-
phologies (Hayes and Kana, 1976). The Rhode Island coast is a
microtidal coast, according to the tidal ranges taken from the
N.O.A.A. Tide Tables. The mean tidal ranges above mean low water vary from 0.76 m at Watch Hill to 1.07 m at Narragansett,
with the extreme spring tidal ranges being 0.94 at Watch Hill
and 1.34 m at Narragansett. The flood tidal deltas on the
Rhode Island coast thus are only intermediately developed, with
poorly developed ebb tidal deltas as suggested by the microtidal
range.

FLOOD TIDAL DELTA SEDIMENTATION ANALYSIS (1939-1975)

The tidal delta at Weekapaug Inlet consists of large subtidal lobes dissected by a bifurcating, sinuous channel and
widespread, fairly continuous supratidal, vegetated deposits.
Over 45,000 m^2 of tidal delta deposits have developed from
subtidal shoals in 1939 to supratidal deposits in 1975 (Figs. 5
and 6, Subtidal and supratidal changes.) The subtidal delta
shoals also extended further into the lagoon in 1975 than they
did in 1939, by the addition of over 73,000 m^2 of sediment. Probably a major factor in effecting these changes was the stabilization of the inlet with jetties and revetments, and straightening and deepening the inlet with dredging in the period between
1951 and 1963. Straightening and dredging of the inlet increased

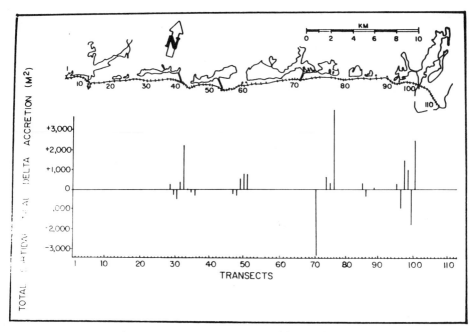

Fig. 5. Total area of subtidal flood tidal delta accretion (m²) during period 1939 to 1975 at the four inlets in the Rhode Island barrier beaches. Areal changes measured photogrammetrically along shoreline segments 300 m wide. Subtidal accretion, in general, is much less effective than supratidal accretion in total landward transport. Segments 1-113 are also erosion survey stations.

the tidal prism of the lagoon, resulting in higher flow velocities with consequently greater amounts of sediment being transported into and eventually deposited in the lagoon.

At Quonochontaug Inlet the subtidal portion of the tidal delta has accreted more than 46,000 m² and the supratidal portion has accreted more than 57,000 m². Some of this change has resulted from the dredge and fill operation associated with the inlet stabilization, but the greater amount of change has probably resulted from the rejuvenation of the tidal prism by the stabilization, as well as the normal trend towards sediment accumulation in the lagoon.

The Charlestown Inlet channel, although straightened somewhat during its stabilization in 1951-1952, remains sinuous and bifurcating towards the distributaries. The overall form of the delta is more like that of the classic delta: essentially triangular or arcuate in plan view, with lobes of subtidal or supratidal deposits dissected by stable to somewhat migratory

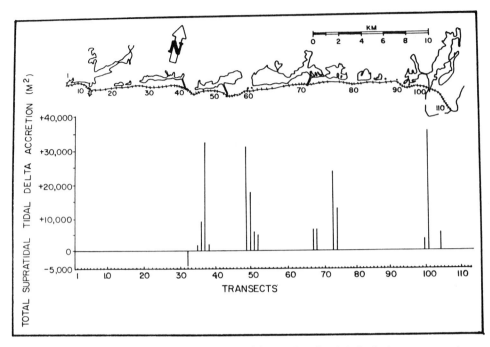

*Fig. 6.Total area of supratidal flood tidal delta accretion
(m²) during period 1939 to 1975 at the four inlets in the Rhode
Island barrier beaches as measured photogrammetrically. Subtidal and supratidal flood tidal delta accretion is one and
one-third times greater than washover accretion along this
same barrier shoreline.*

channels. There are subtidal shoals similar in location within
the tidal channels to fluvial point bars, with the thalweg of
the channel occurring at the outer, or undercut, bank. Most of
these subtidal shoals are nearly intertidal, being almost emergent at spring low tides. Photogrammetric measurements indicate
that more than 432,000 m^2 of subtidal delta shoals have accreted and over 46,000 m^2 of delta deposits were converted from
subtidal shoals in 1939 to supratidal deposits in 1975 in Ninigret Pond. Incorporated in the value denoting change in the
subtidal shoals is a loss of subtidal shoals in the region
north of the Green Hill Pond narrows, in Ninigret Pond. This
loss of subtidal shoals was apparently a result of the channel
dredging that diverted the tidal flow that exits from Green
Hill Pond through the narrows from a northward route around the
glacial islands in the eastern portion of Ninigret Pond, to a
more direct westward route. Before the channel was dredged,
sediment would have been deposited by the Green Hill Pond ebb

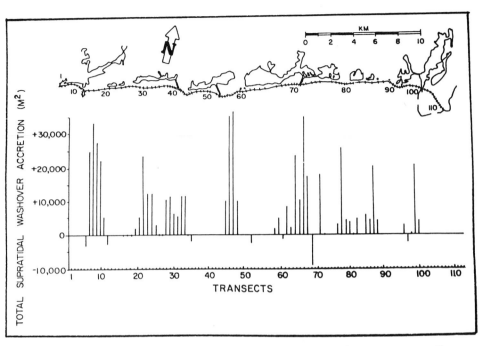

Fig. 7. *Total area of supratidal washover accretion (m²) during the period 1939 to 1975 as measured photogrammetrically along shoreline segment 300 m wide. These measurements indicate that supratidal washover accretion is three times more effective than supratidal delta accretion.*

flow in the area to the north of the narrows in Ninigret Pond. After the channel was dredged and the sediment supply was cut off to this area, minor currents and waves in Ninigret Pond were apparently able to disperse the sediment formerly deposited by the Green Hill Pond ebb flow. The dispersal of the sediment was manifested on the aerial photographs as a decrease in the extent of the subtidal shoals.

The tidal delta at the Point Judith Breachway has undoubtedly been changed greatly in form as a result of the inlet stabilization. The channel has also been dredged, notably between 1951 and 1963, with the dredge spoils being used as land fill to build docks on the eastern bank of the breachway at Galilee and to produce habitable property on the western, Jerusalem bank of the breachway. The tidal delta presently appears as a maze of tidal creeks and subtidal shoals, and intertidal marsh and tidal flat deposits in the regions away from the breachway. Subtidal shoals extend parallel to the main, bifurcating channels as fingers and arcuate lobes, with some intertidal shoals,

mostly mussel and clam flats, developed on the subtidal shoals.
The amount of measured change in the area of the tidal delta
south of Great Island indicates an increase in the supratidal
deposits of nearly 43,000 m^2 and an increase in subtidal delta
deposits of over 198,000 m^2. Much of this change may be mar-
ginally attributable to the effects of dredging and land filling, although an attempt was made to discern changes of an
artificial nature from those of a natural character.

WASHOVER SEDIMENTATION ANALYSIS (1939-1975)

    In general, there have been substantial increases in wash-
over sedimentation, both subtidal and supratidal, during the
36-year period along the entire Rhode Island barrier shoreline.
While sedimentation usually is observed to increase in a regu-
lar pattern, certain depositional variations do occur and significant ones are pointed out in the following summary of the
photogrammetric analysis. Figure 7 indicates the amount of
supratidal washover changes, while Figure 8 indicates that of
subtidal washover changes.
    At the western end of the Rhode Island shoreline, over
107,000 m^2 of supratidal and over 102,000 m^2 of subtidal washover deposits have accreted on the backbarrier of Napatree
Beach spit. On the Maschaug Pond barrier more than 78,000 m^2
of supratidal and nearly 58,000 m^2 of subtidal washover deposits have been deposited, mostly in the western portion of
Maschaug Pond. Along the Winnapaug Pond barrier, over 54,000 m^2
of subtidal deposits have accumulated, again largely in the
western part of the pond where the width of the backbarrier is
at a minimum. Changes in the subtidal shoals in the region of
the 1938 hurricane breach at Misquamicut Beach have been attributed to tidal delta rather than overwash processes, although
overwash has undoubtedly been operative along this narrow section of the barrier beach.
    On the Quonochontaug Pond barrier, over 90,000 m^2 of supra-
tidal and more than 83,000 m^2 of subtidal washover deposits
have accreted, most of this at the western edge of the pond.
To the east, the effects of overwash are masked by the effects
of tidal delta deposition, if overwash is at all operative along
that section of the beach where the dunes are relatively high
and the backbarrier is very wide. On the backbarrier of the
Ninigret Pond barrier more than 86,000 m^2 of supratidal and
more than 50,000 m^2 of subtidal washover deposits have accumulated. A loss of over 54,000 m^2 of subtidal washover deposits in the western portion of the pond is concealed by the
value of total pond-width change. This loss of subtidal washover shoals resulted from supratidal washover accretion

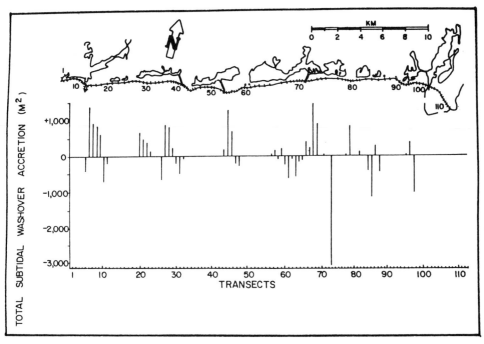

*Fig. 8. Total area of subtidal washover accretion (m²)
during the period 1939 to 1975, as measured photogrammetrically
along shoreline segment 300 m wide. In general, washover accretion, both supratidal and subtidal, is more effective along the
western two-thirds of the Rhode Island coast.*

occurring more rapidly than adjacent subtidal washover deposition and does not represent erosion, but rather the lack of
additional deposition.

Supratidal washover accretion in Green Hill Pond amounted
to more than 31,000 m², while subtidal washover accretion of
nearly 3,100 m² occurred, largely in the eastern portion of the
pond at an isolated washover sluice. At Trustom Pond more than
28,000 m² of supratidal washover deposits accreted on the back-
barrier, with a gain of over 11,000 m² of subtidal washover
shoals. Incorporated in this value is a loss of subtidal deposits which, again, may be attributed partly to the more rapid
accumulation of supratidal deposits and to a less rapid accumulation of adjacent subtidal material. Similarly, a loss in
subtidal washover shoals (more than 10,000 m²) and a minor
amount of supratidal accretion (more than 4,000 m²) occurred in
Cards Pond. This loss of subtidal deposits is essentially identical to that in Trustom Pond, as the form and location of the
transitory inlet into Cards Pond have also changed with time.

A significant amount of washover accretion (nearly 8,000 m^2) occurred in the small coastal pond at Matunuck Point. Just east of this pond more than 20,000 m^2 of supratidal washover accretion has occurred on Potters Pond backbarrier, most of it along the southern edge of the pond. Most of this change occurred as infilling of former tidal channels across the backbarrier by overwashed sediment. Subtidal washover accretion was relatively insignificant because of the predominance of tidal delta accretion changes caused by dredging and land filling in Potters Pond.

WASHOVER SEDIMENTATION AND BARRIER DEVELOPMENT –
ENVIRONMENTAL IMPLICATIONS

The total amounts of areal changes of supratidal and subtidal washover and tidal delta deposits as calculated for the period of 1939 to 1975 is as follows: total areal change of supratidal plus subtidal washover deposits was +522,792 + 267,953 m^2 = +790,745 m^2; total areal change of supratidal plus subtidal tidal delta deposits was +188,238 + 862,322 m^2 = +1,050,560 m^2. Annual rate of areal changes of washover deposits for the whole south shore was calculated to be +21,965 m^2/yr; for annual tidal delta accretion, +29,182 m /yr. According to these values of areal changes, subtidal plus supratidal tidal delta sedimentation is 1 1/3 times more effective than subtidal plus supratidal washover sedimentation in the landward transportation, deposition, and storage of sediment. Supratidal washover accretion, however, is nearly three times more effective than supratidal tidal delta accretion.

A photogrammetric survey of coastal erosion trends along the south shore of Rhode Island over the period 1939-1972 was conducted by Regan and Fisher (1977), with 113 transects spaced about 250-500 m apart along the 40 km of the south shore. Their results indicate that most of the south shore beaches have experienced erosion, particularly in the post-1938 hurricane recovery period between 1939 and 1951. The greatest factor controlling the occurrence and amount of washover accretion during the study period appears to be that of an erosional beach (Fig. 9). At 27% of the transects at which washover accretion was above average (i.e., more than the mean value of +18,000 m^2), beach erosion was also greater than the average of the mean value -6,000 m^2. At least 66% of the transects at which washover accretion was present or above average, beach erosion was also present or greater than the average (Fig. 10). Other factors related to overwash are the height and continuity of the dunes, the development of transitory inlets, and the

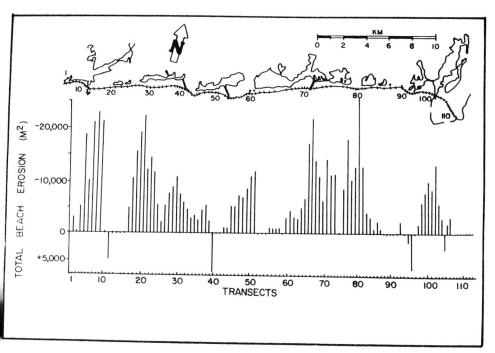

Fig. 9. Total areal amount (m²) of high tide shoreline erosion along the Rhode Island coast for the period 1939 to 1975 as determined photogrammetrically for 113 shoreline segments, 300 m wide. Note that the only beach accretion during this time has been associated with updrift deposition related to groins and jetties.

width of the barrier beach (which is a function of the development of tidal deltas and washover backbarrier deposits and of the amount of beach erosion). All of these features are characteristic of an erosional beach.

Leatherman (1976) monitored overwash along the northern barrier shoreline of Assateague, Maryland during a 26-month interval. It was calculated that 31 percent of the overwash sediment was derived from the beach and dunes during the December, 1974 northeaster. This same storm resulted in the deposition of 20 m³ of overwash material per unit meter length of dune breach (Leatherman, Williams and Fisher, 1977). Earlier, Dolan (1972, 1973) wrote that prevention of overwash by man-made protective foredunes at Cape Hatteras National Seashore on the Outer Banks of North Carolina would have severe consequences. He argued that the material that would have been deposited as washovers was instead lost to the offshore or

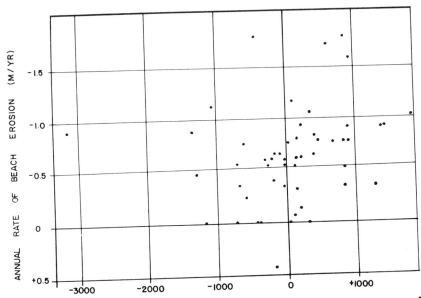

ANNUAL RATES OF SUBTIDAL AND SUPRATIDAL WASHOVER ACCRETION (M²/YR)

Fig. 10. Relationship of annual rate of beach erosion along the Rhode Island barrier shoreline to the annual rates of washover accretion (m²/yr). Both rates determined from photogrammetric measurements for the period 1939 to 1975. Beach erosion was greater than the average of 0.7 m/yr at 27% of the stations at which washover accretion was also above the average; and at 66% of the stations at which this washover accretion was present or above the average, beach erosion was also present or above the average.

carried alongshore. More recently, Boc and Langfelder (1977) measured washovers spatially through time using aerial photographs and showed that 85% of the entire North Carolina coast has experienced overwash since 1938.

Eustatic sea level rise is also probably a factor causing overwash on a worldwide basis. Sea level along the North Carolina coast is rising at a rate greater than the average for the U.S. shoreline (Hicks and Crosby, 1974; Fisher 1977). By contrast, Plum Island, Massachusetts has a stable sea level and landward transgression occurs primarily by dune migration (Cameron and Jones, 1977).

Dillon (1970), from a stratigraphic study of the Charlestown and Green Hill barrier and lagoon sediment, also believed that long-term erosion is the general trend along the Rhode Island south shore. The tendency is for the barrier beaches

to be submerged by the rising sea level. It appears that the
rate of shoreline erosion is greater than would be expected
from only submergence during the past 36 years (Fisher and
Regan, 1977). It is suggested that the increased shoreline
erosion is necessary to restore the offshore profile of equilibrium due to the rising sea level (Bruun 1962). Similar increased erosion during the same time period has been determined
for the North Carolina barrier island coast (Fisher 1977). The
total amount of shoreline erosion is perhaps dependent upon sea
level rise, with most of the eroded shoreline sediment moving
offshore to compensate for the sea level rise and the remainder
being displaced landward as washover and inlet deposits.

ACKNOWLEDGMENT

Funds for this study were provided by the University of
Rhode Island Sea Grant 506-010, Project R/E-9, Principal Inves-
tigator, John J. Fisher, for main study "Photogrammetric Remote
Sensing Inventory of Long Term Rhode Island Shoreline Changes,
Part I - Southern Barrier Coasts; Part II - Narragansett Bay
Coast.

REFERENCES

Bartberger, C.E. (1976). Sediment sources and sedimentation
    rates, Chincoteague Bay, Maryland and Virginia. *J. Sed.
    Petrology 46,* 326-336.
Boc, S.J. and Langfelder, J. (1977). An analysis of beach over-
    wash along North Carolina's coast. Rept. No. 77-9, Center
    for Marine Studies, North Carolina State Univ., Raleigh,
    N.C., 17 p.
Bruun, P. (1962). Sea level rise as a cause of shore erosion.
    J. Waterways and Harbors Div. 88, 117-130, Amer. Soc. Civil
    Engineers Proc.
Caldwell, D.M. (1972). A sedimentological study of an active
    part of a modern tidal delta, Moriches Inlet, Long Island,
    New York. Unpublished M.S. thesis, Columbia Univ., N.Y., 70 p.
Cameron, B., and Jones, J.R. (1977). New evidence for barrier
    island migration at Plum Island, Mass. *In* Proceedings 1976
    New England - St. Lawrence Valley Geog. Soc., Amer. Assoc.
    Geog., v. 6, p. 94-97.
DeAlteris, J.T. (1976). A speculative model for the evolution
    of a barrier island, tidal inlet-lagoon system: I. Regional
    Analysis. Geological Society America Abs. with Programs,
    v. 8, p. 159.

Dillon, W.P. (1970). Submergence effects on a Rhode Island barrier and lagoon and inferences on migration of barriers. *J. Geology 78*, 94-106.

Dolan, R. (1972). The barrier dune systems along the Outer Banks of North Carolina, a reappraisal. *Science 179*, 286-288.

Dolan, R. (1973). Barrier islands, natural and controlled. *In* "Coastal Geomorphology" (D.R. Coates, ed.), p. 263-278. Publications in Geomorphology, S.U.N.Y., Binghamton, N.Y.

Hicks, S.C. and Crosby, J.E. (1974). Trends and variability of yearly mean sea level, 1893-1972. NOAA Tech. Memo. No. 13, Nat'l Ocean Survey, U.S. Dept. Commerce, Rockville, Md., 14p.

Fisher, J.J. (1962). Geomorphic expression of former inlets along the Outer Banks of North Carolina. Unpublished M.S. thesis, Univ. North Carolina, Chapel Hill, 120 p.

Fisher, J.J. (1977). Holocene sea level rise and the Per Bruun theory of shoreline erosion, Middle Atlantic States, Rhode Island and North Carolina. Int'l Quaternary Assoc. Congress, Birmingham, England, Proceedings.

Fisher, J.J. and Regan, D.R. (1977). Relationship of shoreline erosion to eustatic sea level rise, R.I. coast (abs). Northeastern Geol. Soc. Amer., Annual Meeting, Binghamton, N.Y. p. 317.

Fisher, J.S., Leatherman, S.P. and Perry, F.C. (1974). Overwash processes on Assateague Island. 14th Cong. on Coastal Engineering, v. 11, p. 1194-1212.

Godfrey, P.J. (1973). Comparison of ecological and geomorphic interactions between altered and unaltered barrier island systems in North Carolina. *In* "Coastal Geomorphology"(D.R. Coates, ed.), p. 239-258. Publications in Geomorphology, S.U.N.Y., Binghamton, N.Y.

Godfrey, P.J. (1974). The role of overwash and inlet dynamics in the formation of salt marshes on North Carolina barrier islands. *Ecology of Halophytes*. Academic Press, New York, p. 407-427.

Godfrey, P.J. (1976). Supplement to New England Intercollegiate Geol. Conf. Guidebook - Field Trips, Cape Cod, Mass. Dept. of Botany and National Park Service Cooperative Research Unit, Univ. Mass., Amherst, 29 p.

Godfrey, P.J. and Godfrey, M.M. (1975). Some estuarine consequences of barrier island stabilization. *In* "Estuarine Research, Geology and Engineering" (L.E. Cronin, ed.), v. 2, p. 485-516. Academic Press, N.Y.

Hayes, M.O. and Kana, T.W., eds. (1076). Terrigenous clastic depositional environments: some modern examples: A field course sponsored by the Amer. Assoc. Petr. Geologists. Coastal Research Div. Tech. Rpt. No. 11-CRD, Dept. of Geo., Univ. South Carolina, 302 p.

Hoover, R.A. (1969). Physiography and surface sediment facies of a recent tidal delta, Harbor Island, central Texas coast (abs.). Dissertation Abs., v. 29, p. 3790B.

Leatherman, S.P. (1976). Barrier island dynamics; overwash and eolian transport. Proc. 15th Coastal Eng. Conf., Amer. Soc. Civil Engineers, Chap. 114, p. 1958-1974.

Leatherman, S.P. (1977). Interpretation of overwash sedimentary sequences. Geol. Soc. America Abstracts with Program, v. 9, p. 292.

Leatherman, S.P., Williams, A.T., and Fisher, J.S. (1977). Overwash sedimentation associated with a large scale northeaster. *Marine Geology 24*, 109-121.

Lucke, J.B. (1934a). A study of Barnegat Inlet. *Shore and Beach 2*, No. 2, p. 45-94.

Lucke, J.B. (1934b). Tidal inlets: a theory of evolution of lagoon deposits on shorelines of emergence. *J. Geology 42*, 561-584.

Morton, R.A., and Donaldson, A.C. (1973). Sediment distribution and evolution of tidal deltas along a tide dominated shoreline, Wachapreague, Virginia. *Sedimentary Geology 10*, p. 285-299.

Nichols, R.L. and Marston, A.F. (1939). Shoreline changes in Rhode Island produced by the hurricane of September 21, 1938. *Geol. Soc. Amer. Bull. 50*, P. 1357-1370.

Nordquist, R.W. (1972). Origin, development, and facies of a young hurricane washover fan on southern St. Joseph Island, central Texas coast. Unpub. M.S. thesis, Univ. Texas at Austin.

Pierce, J.W. (1970). Tidal inlets and washover fans. *J. Geology 78*, 230-234.

Regan, D.R. and Fisher, J.J. (1977). Photogrammetric study of Rhode Island coastal erosion. Amer. Assoc. Petroleum Geologists, Annual Meeting, Washington, D.C.

Schwartz, R.K. (1975). Nature and genesis of some storm overwash deposits. U.S. Army Corps of Engineers, Coastal Eng. Research Center, Tech. Memo. No. 61, 69 p.

Scott, A.J., Hoover, R.A., and McGowen, J.H. (1969). Effects of Hurricane "Beulah," 1967, on Texas coastal lagoons. *In* "Coastal lagoons, a symposium"(A. Castanares and F.B. Phleger, eds.), p. 221-236.

Shepard, F.P. (1952). Revised nomenclature for depositional coastal features.*Amer. Assoc. Petroleum Geologists Bull. 36*, p. 1902-1912.

Stafford, D.B. (1971). An aerial photographic technique for beach erosion surveys in North Carolina. U.S. Army, Corps of Engineers, Coastal Eng. Research Center, Tech. Memo. No. 36, 115 p.

Stafford, D.B. and Langfelder, L.J. (1972). Air photo survey
    of coastal erosion. *Photogram. Engr. 37*, p. 565-575.
Stirewalt, G.L. and Ingram, R.L. (1974). Aerial photographic
    study of shoreline erosion and deposition, Pamlico Sound,
    North Carolina. Univ. N.C. Sea Grant Program, Pub. No.
    UNC-SG-74-09, 66 p.
U.S. Army, Beach Erosion Board (1949). South shore, state of
    Rhode Island, beach-erosion control study. House Doc. No.
    490, 81st Congress, 2nd Session.

PROCESSES AND MORPHOLOGIC EVOLUTION OF AN ESTUARINE AND COASTAL BARRIER SYSTEM

John C. Kraft

Department of Geology
University of Delaware
Newark, Delaware

Elizabeth A. Allen

The Shell Oil Company
New Orleans, Louisiana

Daniel F. Belknap
Chacko J. John
Evelyn M. Maurmeyer

Department of Geology
University of Delaware
Newark, Delaware

Atlantic coastal barrier systems vary greatly in morphology and internal structures. Major factors determining evolution of barrier systems are: nearshore surface and submarine morphology; source and size of sediment available; wind, wave, and tidal processes impinging on the barrier system; climate of the region; and variations in relative sea-level change. Coastal barriers in Delaware are rapidly transgressing. Wave-dominated processes are the most important single factor in determining the volume of sediment in motion. Availability of sediment from the continental shelf, erosion of the barrier itself, and erosion of headlands determines whether or not a barrier can exist and evolve.

Studies of the internal structure of the barriers indicate that washover processes and flood tidal delta deposition are the dominant factors in the landward transgression. Wind transported sand is of secondary importance, derived from the beachberm and washover fans and transported to coast parallel dunes.

Coastal barriers vary in width from 4 kilometers (including the submerged portions of the barriers) to thin sandy strands less than 1/4 meter thick and ten meters wide. In some areas of the estuarine system barriers are totally absent due to lack of sand supply and marshes from the shoreline edge. Because of flood tidal delta deposition and overwash, the landward side of the barrier interfingers with lagoonal and salt marsh organic muds. The evolving transgressive barrier-lagoon-marsh system has a definitive set of characteristics which can be identified in vertical sequences depending upon geographic setting. These vertical sequences should allow identification of ancient transgressive barrier systems in outcrops or cores.

INTRODUCTION

   Evolution of coastal barrier systems depends on a wide variety of physical processes acting on the coastal zone as well as the geomorphologic and geologic setting of the coast. Wind, wave, and tidal processes impinging on the coastal zone, regional climate and variations of relative sea level as well as sediment compaction and tectonic uplift or subsidence play major roles in determining the morphology of a coastal system. In addition, source, size, and volume of sediment available for transport and deposition in the coastal zone are of primary importance. Finally, the geomorphology and geology of the region ultimately determine sediment availability as well as shape and evolution of coastal units.
   In the northeast United States, the coastal zone has been strongly affected by climatic and tectonic events resulting from the latest Wisconsin glaciation and waning of the ice sheet to its present position in Greenland and the Arctic Ocean. Thus, the area is one of major interplay over a short geologic time of isostatic, eustatic, climatologic, and morphologic evolution that has led to highly varied coastal forms, from the sandy mid-Atlantic coastal region to the highly irregular form of the rocky coasts of eastern Maine. Throughout the region, both transgression and regression have occurred during the Quaternary Epoch. However, the dominant factor for the northeast coast of North America during the Holocene Epoch has been that of a marine transgression.
   Delaware's coastal zone is being inundated by relative sea-level rise accompanied by coastal erosion, leading to highly varied forms of transgressive shoreline environments. Geomorphology of the land surface undergoing transgression is critical in determining the types of coastal zone stratigraphic units presently forming and migrating landward (Figure 1).

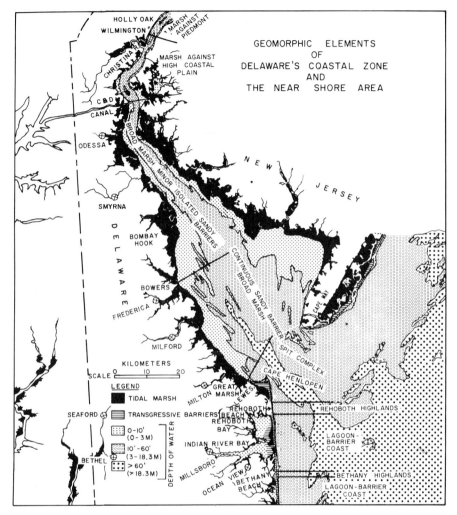

Fig. 1. Geomorphic elements of Delaware's coastal zone and nearshore area relating shore zone configuration to geography and nearshore submarine topography.

Along the Atlantic coast of Delaware, a low-lying coastal plain of probably marine origin during the Sangamon Age is undergoing erosion and transgression. The land surface at the ocean interface includes a lagoon-barrier coast and intermittent highlands with elevations up to 7 meters above sea level. The Atlantic coastal zone includes the highland-embayed barrier shoreline and northerly spit component as identified in the compartmentalization models of Fisher (1967). In addition, the

ancestral valley of the Delaware River of late Wisconsin glacial
time is now a major estuary (Delaware Bay), which extends far
inland, ultimately forming the boundary between the Piedmont
geomorphic province and the northwestern edge of the Mesozoic-
Cenozoic coastal plain of the eastern continental margin of
North America. The Delaware coast may, therefore, be divided
into zones based on the nature of the transgression. Resultant
sedimentary depositional units in the transgressive sequence
are accordingly highly variable over the relatively short (200
km) coastal zone. To the north, a narrow tidal river marsh
fringes crystalline rocks of the Piedmont Province. Farther
south a highly irregularly shaped marsh, infilling relatively
deeply incised pre-Holocene river valleys, is transgressing
across a high coastal plain comprised of Mesozoic and Cenozoic
sediments. Along the lower tidal Delaware River and the northern portion of Delaware Bay, broad marshes and minor isolated
sandy barriers form the shoreline along the relatively low-
lying undulating coastal plain. Along the southern part of
Delaware Bay, a continuous sandy barrier and broad marsh is
transgressing landward across the low-lying coastal plain. The
Cape Henlopen region at the confluence of the Delaware estuary
and Atlantic Ocean includes a regressive sequence of sandy barriers, high dunes, coastal marshes and shallow estuarine sediments. Along the Atlantic coast a lagoon barrier coast is developed between low-lying eroding highlands of the lower Delaware coastal plain (Fig. 1) (Kraft 1971; Kraft and others,
1973; Kraft and Belknap, 1975; Kraft and John, 1976; Kraft and
others, 1976; John 1977; Allen 1977).

The purpose of this study is to develop a set of models of
the varied elements of the coastal zone and equate these models
to physical processes responsible for the development of coastal environments within a relatively small but highly varied
geomorphic terrain undergoing transgression.

PROCESSES INFLUENCING EVOLUTION
OF THE DELAWARE COASTAL ZONE

Relative sea-level rise throughout the Holocene Epoch has
been the dominant factor in coastal transgression (Fig. 2).
During peak late Wisconsin glaciation, a deeply incised topography including the ancestral Delaware River and a trellis-
type drainage system evolved in the area of Delaware's coastal
zone and the Delaware Estuary. Accordingly, with waning of the
Wisconsin ice sheet and subsequent eustatic sea-level rise,
coastal environments migrated landward across the now submerged
continental shelf. At the same time, the deeply incised pattern of the region allowed tidal waters to intrude from one to

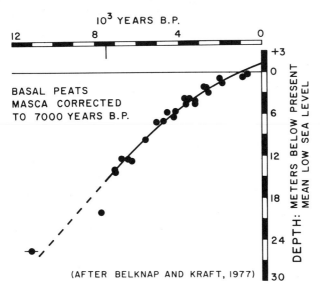

Fig. 2. Local relative sea-level rise curve based on basal organic salt-marsh sediments in the Delaware coastal zone.

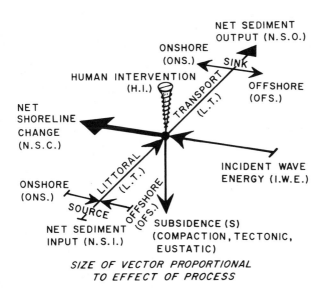

Fig. 3. A composite vector diagram showing processes in action and elements of coastal change in the Delaware coastal zone. The model is conceptual and semi-quantitative as used in this paper.

two hundred kilometers landward since the beginning of the Holocene transgression. As this irregular-trellis drainage system was inundated by eustatic sea-level rise, a highly irregular set of coastal environments developed. The form of these sedimentary bodies and geomorphology of the region remained fairly constant as the coastal zone migrated landward. Initially, transgression was fairly rapid at a rate of 30 cm/century as indicated by a local relative sea-level curve developed in the Delaware coastal zone (Fig. 2). Gradually the rate of relative sea-level rise began to decrease to 20 cm/century approximately 5000 years B.P. At this point coastal erosion became a more important factor in the migration of the coastal environments. In addition, the outer continental shelf began to subside (Belknap and Kraft, 1977; and Kraft 1976).

Figure 3 presents a conceptual vector diagram of important short-term processes of coastal change involved in the evolution of the transgressive forms of Delaware's coastal zone. The vectors shown are measures of relative magnitude shown in a Cartesian coordinate system. The axis is a measure of longshore sediment transport budget, and includes both input into the system by erosion and longshore transport and output by deposition and longshore transport. With the presence of a highland, there is a net gain of sediment to the littoral transport system, which is carried alongshore in the nearshore shallow marine areas. The Y-axis is a measure of onshore-offshore components. Incident wave energy is the dominant element for landward erosion and migration of the coast, resulting in net landward shoreline retreat. Important processes of this transgression include high volume offshore sediment transport (leaving material permanently in the inner marine nearshore area), and storm overwash, an important element of landward barrier migration. The Z-axis represents human intervention, which is important in some parts of the coastal zone, as it results in alteration of natural processes. Also plotted on the Z-axis is relative subsidence. Relative subsidence includes eustatic sea-level change, tectonic subsidence, local compaction and possibly glacial isostatic effects (Belknap and Kraft, 1977).

MODELS FOR THE EVOLUTION OF COASTAL BARRIER SYSTEMS

Analysis of geomorphology, subsurface geology, coastal processes, and paleogeographic reconstructions suggests that the Delaware coastal zone can be represented by several models. These include (1) an Atlantic coastal lagoon transgressive barrier system dominated by tidal inlets; (2) an Atlantic coastal lagoon-linear washover barrier and eroding pre-Holocene headland in which longshore transport and washover processes are dominant;

(3) a complex spit system evolving at the confluence of the
Delaware Estuary and the Atlantic Ocean; (4) an estuarine washover barrier-broad marsh system under the influence of tidal
and wave action of Delaware Bay; and (5) an area of marine
transgression across broad coastal marshes along the lower tidal
Delaware River.

Atlantic Barrier - Tidal Inlet Dominated

In a transgressive lagoon barrier coastal system, flood tidal deltas tend to be dominant over ebb-tidal deltas. This is
primarily because of the development of flood-tidal deltas in
lower energy lagoon environments, whereas ebb-tidal deltas are
developed in the nearshore marine area, where there is extremely
high wave activity and longshore currents. Along the axes of
ancestral stream valleys, sequences of sediments up to 30 meters thick are deposited. Where highlands crop out at the
shoreline, deposition is low and erosion is dominant. Tidal
inlets are subject to opening by storm processes and tend to
migrate before closing caused by sedimentation due to wave activity and longshore transport.

Along the Atlantic coast of Delaware at least three and
possibly five major tidal deltas have formed over an 8 km coastal segment since the mid-17th century. Figure 4 is an oblique
aerial photograph of the Atlantic Ocean barrier with the Atlantic Ocean on the left (east) and coastal lagoons (Indian River
and Rehoboth Bays) on the right (west). As flood-tidal deltas
stabilize and inlets are filled, the inlet sequence migrates
and the older flood-tidal deltas tend to stabilize. Back-
barrier marshes develop upon the tidal flats between tidal
channels of the abandoned inlet (Fig. 4). As transgression continues, sediment is moved through the system by longshore transport and storm overwash from "northeasters" and occasional hurricanes. Storm overwash processes lead to deposition of washover fans over the back-barrier marsh. Major storms cause deposition of very large washover fans which extend into adjacent
coastal lagoons.

The sandy Atlantic coast barrier (Fig. 5) includes thin,
narrow coast-parallel dunes, a beach-berm system, washover fans,
flood-tidal deltas, nearshore marine offshore bars and possibly
small ebb-tidal deltas. A thin nearshore marine sand and gravel unit overlies the transgressed sequence. Tidal delta sands
interfinger with lagoonal muds in actively building tidal deltas. Interfingering of back-barrier marsh sequences with washover sediments is common (Fig. 6). Interfingering of flood-
tidal delta sands with lagoon-marsh muds may or may not develop,
depending on geomorphic configuration. Thus, in subsurface or
outcrop, it should be possible to identify the landward side of

*Fig. 4. An oblique aerial photograph of the Atlantic
coastal barrier in an area of modern and ancient delta
and inlet formation.*

a transgressive barrier system with flood tidal deltas and/or
washover fans.

Wave erosion and net shoreline recession are dominant as
transgression continues landward. Littoral transport is an
important process of moving sediment laterally along the coast.
In this way some of the sediment is moved out of the system to
adjacent coastal areas, and some sediment is permanently lost
offshore. The net movement of sediment in a landward direction
maintains configuration of the barrier and its subenvironments.
Sedimentary lithosomes developed (Fig. 5) are thus growing in
a landward and upward direction and being truncated along the
foreshore-shallow marine face.

In the case illustrated in Figure 5, human intervention is
of major importance in that efforts have been made to stabilize
dunes to protect the barrier from overwash. Efforts such as
this are believed to lead to accelerated erosion of the berm-
beach-foreshore during storm wave activity (Dolan 1973). Artificial stabilization of Indian River Inlet by jetties has resulted in sand accretion on the south (updrift) side and major
erosion on the north (downdrift) side, endangering a coastal
highway which is parallel to the shoreline.

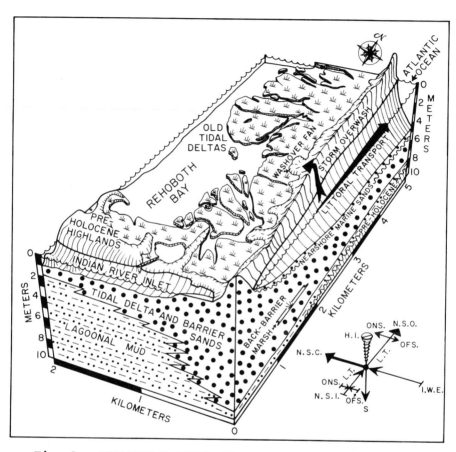

*Fig. 5.   ATLANTIC BARRIER AND TIDAL DELTAS.   A block dia-
gram showing surface geomorphic and subsurface stratigraphic
elements of an Atlantic coastal barrier in the vicinity of
tidal deltas.   See Figure 3 for interpretation of vector
diagram.*

Preservation of the transgressive sedimentary sequence depends on rate of relative sea-level rise rather than rate of
development of transgressive sedimentary units and their landward migration (Fischer 1961).   Thus, there would be a tendency to preserve the landward and lower elements of the transgressive sequences of this form of the transgressive barrier
model.

Fig. 6. A photograph of a trench in the washover zone on the back of an Atlantic coastal barrier showing two sandy washover units separated by in situ marsh sediments. A modern high-marsh flora is presently extant at the top of the photo.

Linear Atlantic Baymouth Barrier and Pre-Holocene Headland

The linear baymouth barrier headland system such as that near Dewey Beach and Rehoboth Beach, Delaware (Fig. 7), is subject to processes similar to those of the previous model of a baymouth barrier-tidal delta system. A linear baymouth barrier is developed mainly by longshore transport and storm overwash processes. Tidal inlets and associated flood-ebb tidal deltas are absent. As before, the pre-existing topography being transgressed is a controlling factor determining the thickness and shape of the sedimentary environmental sequences formed during transgression. The baymouth barrier and it subenvironments (Fig. 8) extend landward and interfinger with back-barrier marshes and lagoonal muds. In the beach-berm-foreshore area, erosion occurs during times of storm wave activity. A considerable amount of sediment is transported alongshore in this

Fig. 7. An oblique aerial photograph of a linear baymouth barrier looking from Rehoboth Bay toward the barrier and Atlantic Ocean to the right and the Rehoboth highland to the north. Compare with block diagram shown in Figure 8.

area. Turner (1968) estimated that net longshore transport to the north along the Atlantic coast of Delaware varies from 100,000 to 340,000 cubic meters per year.

Proof of landward transgression of the linear washover barrier is provided by frequent exposure of the back-barrier tidal marsh in the swash zone at low tide after a storm. In some cases a back-barrier pine forest is exposed when the 2 m thick berm is temporarily removed after a storm. Radiocarbon dates on this pine forest average about 300 years before present and thus give a measure of the rate of transgression in the short-term past. Subsurface borings frequently encounter back-barrier marshes as thin lenses of marsh muds within barrier sands. These are also an indicator of dominant landward and upward migration of the linear barrier throughout the transgression.

The vector diagram in Figure 8 shows that incident wave energy is the major cause of net landward migration of the barrier. In the presence of a highland, there is net input of sediment to the littoral transport system. Relative sea-level rise is the same throughout coastal Delaware, with minor

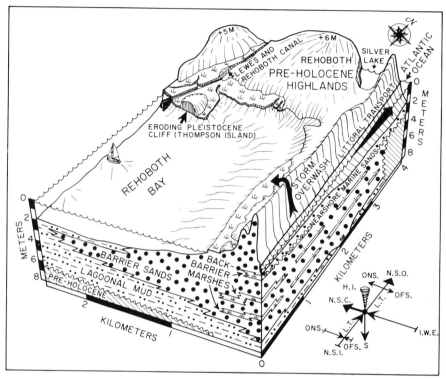

Fig. 8. ATLANTIC BAYMOUTH BARRIER AND HEADLAND. A block diagram of a linear Atlantic baymouth barrier and headland showing surface morphologic features and sub-surface sedimentary unit distribution. The vector diagram shows a semi-quantitative estimation of various processes in action as the coastal environments transgress. See Figure 3 for interpretation of the vector diagram.

variance along the coast in areas of deeply incised pre-Holocene valleys where sedimentary compaction adds an extra component of subsidence.

Figure 9 presents a vertical sedimentary sequence and photographs of a core drilled through the linear baymouth barrier. Important elements to note are the sequence (from top to bottom) of dune, barrier, back-barrier marsh, and lagoonal-fringing marsh sediments. These overlie the unconformable pre-Holocene transgressive surface which in this case is a surface incised into Sangamon (?) Age marine sediments. The vertical sedimentary sequence is equivalent to the lateral sedimentary environmental sequence shown in the block diagram in Figure 8. In this and other similar geological studies of the coastal zone of

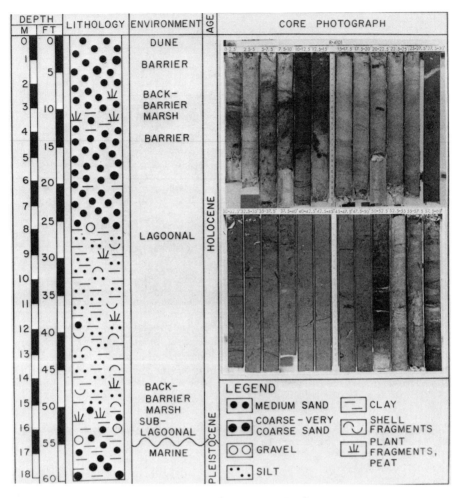

Fig. 9. *A lithologic description of the vertical sediment sequences encountered in a linear baymouth barrier , Atlantic coastal Delaware, showing environments of deposition and a core photograph of the Holocene environmental sequences. Barrier sands extend from 0 to -3 feet. From -3 feet to -10 feet high-marsh floras have colonized the washover sands. A back-barrier marsh is developed from -10 feet to approximately -12 feet. From approximately -12 feet to -27.5 feet sub-lagoonal sands occur. From -27.5 feet to approximately -40 feet lagoonal muds including* Crassostrea virginica *occur. From approximately -40 feet to -52.5 feet a mixture of lagoonal muds and fringing salt-marsh muds occur, terminating at a basal peat. Below these Holocene sediments occur similar coastal environmental sediments of Sangamon Age, strongly oxidized (core photographs courtesy of Shell Development Company, Houston, Texas).*

Fig. 10. An oblique aerial photograph of the Cape Henlopen spit complex looking toward the recurved spit tips in the fore-ground, the Atlantic Ocean to the far right, and the simple spit Cape Henlopen in the top of the photograph. Compare with the block diagram Figure 11.

Delaware, the vertical sedimentary sequence provides an example of Walther's Law of Correlation of Sedimentary Facies (Kraft 1971; John 1977).

Ocean-Estuarine Spit System

North of the Rehoboth highland, extending to Cape Henlopen and westward to the town of Lewes along the southern part of Delaware Bay, lies a triangular tract of land which is a com-plex spit system. This spit system is the continuation of a foreland between the Delaware Estuary and the Atlantic Ocean which migrated landward from a former position on the Atlantic continental shelf (Fig. 10). Essentially, this system is eroding along the Atlantic shoreline and accreting at the tip of the spit, Cape Henlopen, and along the Delaware estuarine shoreline in the area of Lewes Harbor. Very detailed studies have been made of this area (Kraft and others, 1978; John 1977; Allen 1974; Maurmeyer 1974; and Demarest 1978). By means of surface morphologic analysis, drill-hole data and study of

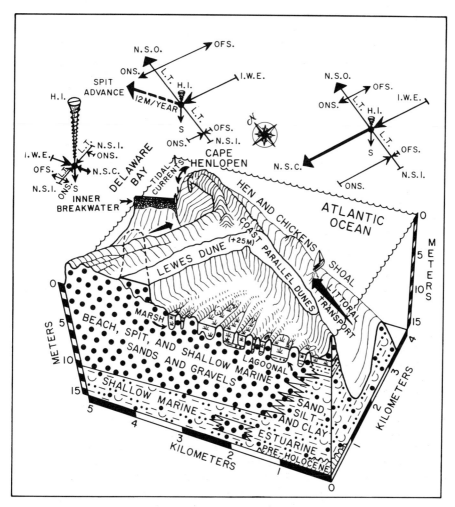

Fig. 11.   OCEAN-ESTUARINE SPIT SYSTEM.   A block diagram of an ocean-estuarine spit system, Cape Henlopen.   Surface morphologic elements are compared with interpretation of subsurface distribution   of beach, spit and shallow marine sands and gravels as contrasted with estuarine-shallow marine sediments that are being covered by the advancing spit complex.   Three vector diagrams are shown to illustrate the highly varied processes that impact on the triangular shaped spit complex as it evolves and transgresses landward along the Atlantic coast and builds bayward along the Delaware Estuary shoreline.   Compare the vector diagrams with Figure 3 for key to the symbols.

coastal erosional and depositional processes, it has been established that Cape Henlopen first evolved from a recurved spit
system to a cuspate foreland to the present simple spit, which
is advancing rapidly into Delaware Bay (Kraft and others, 1978).
In the early stages of spit evolution, when recurved spits developed in the area, they gradually cut off an embayment at the
southern edge of Delaware Bay, which became an isolated lagoon.
This later filled with sediments and then became a tidal flat
on which a broad salt marsh comprised mainly of *Spartina alterniflora* developed (Allen 1974).

A block diagram of the ocean-estuarine spit system in the
Cape Henlopen area is shown in Figure 11. The complex spit
system is essentially a regressive sequence within the overall
transgression of the Delaware coast, as spit sands overlie
shallow marine-estuarine sediments of earlier Holocene Age.
The spit system is at the northern end of a longshore transport
system carrying sediment northward along the Atlantic coast of
Delaware. Accordingly, an excess of sand and gravel is transported to the area. Some of the sediment transported to the
tip of Cape Henlopen is removed by tidal currents to an ebb-
tidal shoal (Hen and Chickens Shoal) which extends over 10 km
to the southeast onto the nearshore continental shelf. The
area is subject to winds from the northwest, north, east, and
southeast across the beaches and tidal flats. Thus, the coast-
parallel dunes of the Atlantic Ocean developed to elevations
of 10 m, whereas a large dune (Lewes dune) with elevations up
to 25 m above sea level has developed parallel to the estuary
shoreline. The coast parallel dunes are migrating landward
across the recurved spit tips at a fairly rapid rate. The
Lewes Dune is migrating southward in response to the prevailing
northwesterly wind direction in the area.

In the early 19th century a breakwater was constructed in
the Lewes Harbor area and the wave and sediment transport regime was drastically altered. Cape Henlopen was transformed
rapidly from a cuspate foreland to a simple spit migrating into
deep water at the mouth of the Delaware Estuary. The Lewes
Harbor area gradually became dominated by deposition. Silt and
mud were deposited in the harbor and sands in the shallow tidal
flats to the west and north of Cape Henlopen. Thus, man's intrusion into the area has had a major effect. An analysis of
historical events leading to the transformation from a cuspate
foreland to a simple spit was presented by Kraft and Caulk
(1972). Demarest (1978) determined the volume and rates of
sedimentation and erosion in the harbor system as related to
the construction of breakwaters and alteration of tidal current
and littoral transport flow patterns. Maurmeyer (1974) developed a detailed compilation of erosion rates on the Atlantic
coast of Cape Henlopen and depositional rates on the spit tip

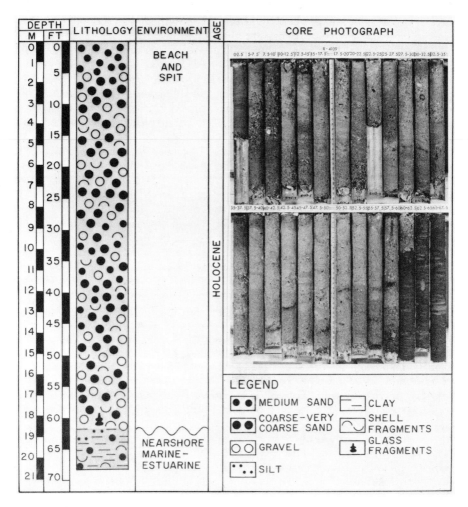

Fig. 12. A lithologic description of a core drilled at the tip of Cape Henlopen in 1964 showing environments of deposition and a core photograph. The first 2½ feet of the core photographed are the surficial beach sands of the spit. From 2½ feet to 60.5 feet the poo·ly sorted sands and gravels of the spit occur with high angle cross-lamination. Under the conformity at 60.5 feet occurs a shallow marine-estuarine sequence of thinly interbedded sands and muds with intense bioturbation. These sands and muds are of Holocene Age and continue into the shallow lower Delaware Estuary immediately adjacent to the point of spit growth. (Core photographs courtesy of the Shell Development Company, Houston, Texas.).

as it rapidly accelerated in rate of advance (up to 30 meters per year at the present time).

As can be seen from Figure 11, vector diagrams of coastal processes in the spit complex are highly variable. The rate of landward erosion is up to 3 meters per year, averaged over the past two centuries. This mass of sediment is moved landward into coastal dunes and northward to the tip of the spit by longshore transport, thus providing sediment for the rapid rate of spit growth. Therefore, net loss of sediment occurs on the Atlantic Ocean side. It is also an area of major loss of sediment into the Hen and Chickens Shoal and nearshore continental shelf. A vector diagram at the tip of Cape Henlopen shows a major spit movement at a rate of approximately 12 meters per year (averaged over the past 150 years) to the northwest. Littoral transport is important and a large volume of sediment is deposited at the tip of the spit in waters up to 20 meters deep. Some of the sediment moves by longshore transport and tidal currents around the spit and into Lewes Harbor, tending to fill in the eastern end of the harbor with coarser sediments.

Due to human interference, vector diagrams of active processes in Lewes Harbor are very complex. With restriction of wave energy by Cape Henlopen on the east, the inner breakwater to the north, and a jetty to the west of the system, the shoreline exhibits highly variable patterns of erosion and deposition. The harbor itself is filling very rapidly by deposition of mud and silt in the relatively quiet waters behind the breakwater. Large quantities of sand move into the harbor via littoral drift around a jetty to the west and around Cape Henlopen to the east.

Figure 12 presents a vertical sedimentary sequence based on a core drilled on the tip of Cape Henlopen in 1964. Spit sediments attain a thickness of 18 meters and consist of coarse to medium sands and gravels mixed with shells which unconformably overlie nearshore marine-estuarine sands and muds of lower Delaware Bay. Immediately adjacent to the tip of the Cape, the ancestral Delaware River valley was incised to a depth of greater than 70 m below present sea level. Estuarine and shallow nearshore marine sediments are infilling a portion of this valley. Portions of the spit complex at Cape Henlopen are filling relatively deeply incised tributary valleys of the ancestral Delaware River. Thus, the potential for preservation of the stratigraphic section is great in this area. Potentials for misidentification of environmental sequences in paleogeography are also great when a major "regressive" environmental sequence is intimately linked with an overwhelmingly transgressive sequence.

Fig. 13. An oblique aerial photograph of the continuous estuarine washover barrier between Delaware Bay on the right and the broad coastal marshes of the lower Delaware Estuary.

Estuarine Washover Barrier Marsh System

The shoreline of lower and middle western Delaware Bay includes broad coastal salt marshes with relatively thin and narrow washover barriers and broad tidal flats (Fig. 13). In local areas erosion of low-lying highlands of pre-Holocene sediments of the coastal plain which crop out at the shoreline provides sediment to the system. Deposition of muds in tidal salt marshes is a dominant process at the leading edge of the transgression. Broad tidal flats formed on relict marsh surfaces dissipate wave energy in the nearshore zone. The maintenance of a very thin (1-2 m) and relatively narrow (50 m) barrier over a long distance appears somewhat anomalous. A balance of sediment supply, longshore transport, and overwash processes is critical to the maintenance of a barrier in an estuarine-salt marsh environment. Figure 14 shows a model of the thin estuarine washover barrier. Salt-marsh elements on high, low brackish marshes may be seen in the surface and correlated into the subsurface. The estuarine washover barrier itself is migrating landward at relatively rapid rates averaging 2-3 meters per year as the transgression continues. The barrier migrates by beachface erosion and storm overwash. In a few cases,

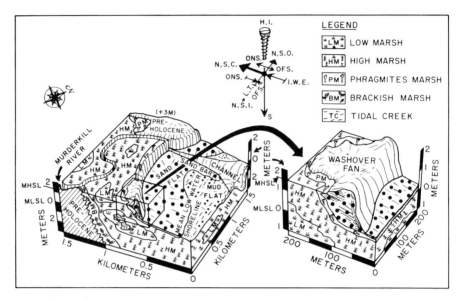

*Fig. 14. ESTUARINE WASHOVER BARRIER, MARSH, TIDAL RIVER,
AND HEADLAND. A block diagram of an estuarine washover barrier,
marsh, tidal river and headland sequence along the southwestern
coast of Delaware Bay. Elements of marsh flora are shown in
surface living position and in the subsurface. An expanded
block of the thin washover barrier including a washover fan is
shown. Erosion is occurring on the beach face with deposition
of semihorizontal washover laminae with high-angle foreset beds
in a landward direction. A sand-mud tidal flat occurs at low
tide in front of the barrier. The vector diagram shows importance of processes in action along a shoreline of this type.
Compare with Figure 3 for key to the vector diagram.*

low coastal dunes are developed on the barrier. However, essentially the entire barrier along Delaware Bay is overwashed in
even modest northeasterly storm conditions or spring high tides.
Near the mouth of the estuary, the barriers increase in elevation, width, and thickness, and are less frequently overwashed.
Towards the north the barriers tend to become narrower and
thinner until the point is reached where they become discontinuous and absent (Fig. 1).

The continuous sand-gravel estuarine washover barrier of
lower Delaware Bay has a relatively complex history. Over the
past several hundred years, a large amount of sand eroded from
the Atlantic coast moved and formed a spit which grew westward
parallel to the shoreline of lower Delaware Bay. Accordingly,
the barrier in this area was higher and wider than to the north
and west, due to a greater supply of sand. These barriers

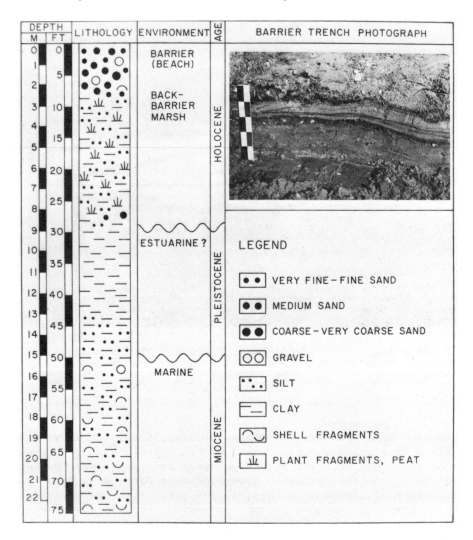

Fig. 15. A lithologic profile of the vertical sequence through a thin estuarine washover barrier and back-barrier marsh. The photograph shows a trench through the poorly sorted subhorizontal washover layers.

consist of a shallow marine foreshore of fine sand and silt, a muddy intertidal flat, a wide sandy barrier (∿100–150 m) with a well-developed dune field, and a gently landward dipping back-barrier washover terrace. The relatively high dune elevations in the southeasterly part of the shoreline area prevented back-barriers from being overwashed in all but the most severe

Fig. 16. An oblique aerial photograph of a linear estuarine washover barrier, showing a broad tidal flat exposed in front of the barrier. The tidal flat includes a major erosional surface of salt marsh sediments.

storms. Shore parallel spit growth during the 19th century resulted in a seaward displacement of the shoreline with an average accretion of three meters per year over a 120 year period.
However, with the northward growth of Cape Henlopen and the construction of breakwaters, groins, and jetties in Lewes, longshore transport no longer supplies sediment to the lower Delaware Bay beaches and the current trend is transgression and
erosion at rates comparable to the rest of the Delaware Bay
shoreline (2-3 m per year). Along the lower part of the Delaware estuarine beaches, net longshore transport is approximately
8,000 to 10,000 m³ per year toward the southeast. A northwest-
ward component of longshore transport occurs primarily during
the summer months when dominant wave approach is from the southeast.
      Grain sizes on the Delaware Bay washover barriers are sig-
nificantly larger than those along the Atlantic coast, having
an average size range of 0 to 1.0 ∅ (coarse sand). This includes a large number of pebbles derived from erosion of Pleistocene fluvial-glacial deposits. Sediments are not as well-
sorted as the Atlantic coastal sands.

Interbedded washover sand-marsh mud sequences are noted in cores along the Delaware Bay coast (Fig. 15). As overwash occurs frequently, there is often a development of a marsh on the washover sand units, resulting in thin laminae or layers of organic debris.

Eroding marsh surfaces (Fig. 16) supply silt and clay and organic debris to the system, making waters of Delaware Bay highly turbid. In effect, the barrier is constantly being re-cycled landward while the muds of the coastal tidal marsh area are recycled into the waters of Delaware Bay where deposition occurs in the nearshore estuarine zone and across the tidal marsh at spring flood tide.

Man's intrusion into the environment is evident along the washover barriers of Delaware Bay. In areas where small coastal resort villages have been developed, groin fields have been constructed. At Broadkill Beach (northwest of Lewes) the groin field has held the beach in position, while to the northwest and southeast the transgression continues, thus tending to place the town of Broadkill Beach in an "island" setting sur-rounded by estuarine waters and a tidal marsh. Sand is fre-quently dredged from offshore shoals and dumped onto the beach to increase sand supply at a rate roughly equal to the rate at which it is being eroded. Thus, in the areas developed by man, beaches tend to be slightly wider and higher, but only so long as man continues to nourish them artificially.

A vector diagram (Fig. 14) shows that human intrusion is relatively high, and relative sea-level rise is about the same along the Atlantic coast of Delaware. Net onshore-offshore movement of sediment is less, and the longshore transport sys-tem moves sediment in quantities approximately one order of magnitude less than along the Atlantic shore. Incident wave energy is much lower in view of the broad, shallow and partly sheltered Delaware Bay. Net shoreline change is highly variable ranging from 1 to 7 meters per year averaged over the past cen-tury. Figure 15 shows a vertical sedimentary sequence of a thin estuarine barrier of sand-gravel and shell debris overlying a marsh sequence which unconformably overlies Pleistocene sedi-ments of estuarine and marine origin. A photograph of a trench in the barrier (Fig. 15) shows sub-horizontal laminations and coarse, poorly sorted sediments of the washover barrier.

The broad salt marshes paralleling the coasts of the Dela-ware Estuary (Fig. 16) have been studied and analyzed in three dimensions by Allen (1977). Although many previous studies have been made of the salt marshes of Delaware, the internal facies were interpreted with considerable question. Allen (1977) applied microtome sectioning techniques to precisely identify decaying organic debris in marsh sediments. With this technique, it became possible to identify various marsh facies such as high marsh, marsh pond, brackish marsh, low marsh,

tidal creek, detrital organic marsh units, etc. (Fig. 17). This
type of analysis enables one to quantitatively establish paleogeography in salt-marsh systems.   Reasons for variants of salt
marsh floral types are numerous.   They include the nature of
the surface undergoing transgression, frequency and closeness
of fresh-water drainage systems to the edge of the salt marsh,
microtopography of the marsh surface, nature of the barriers
between the marsh and the estuarine system, number of tidal inlets or breakthroughs of the barrier, nature of marsh drainage
patterns, and many other elements.   By determining precise
floral elements in place in marsh cores, a pattern of paleogeographic settings of marshes can be put together and accordingly, typical vertical sequences can be determined for different shoreline types.   Figure 17 illustrates three typical vertical sequences of marsh sediments representing shoreline deposition along a broad marsh-estuarine-barrier setting.   With
the help of radiocarbon dating, it is possible to project time-
lines through interpretations such as that shown in Figure 17.
Thus, paleogeography of the marsh surface can be identified
and projections can be made of positions of adjacent highlands,
tidal creeks, marsh facies, type of barrier and presence or
absence of tidal inlets.

Four major types of marsh settings along the western shore
of Delaware Bay can be identified.   These include the low marsh
stands of *Spartina alterniflora* which occur within the tidal
zone and fringe tidal creeks; high marsh settings of *Distichlis
spicata* and *Spartina patens; Phragmites communis* stands, which
frequently occur around marsh ponds or colonize the backs of
estuarine barriers; and high marsh or brackish pond systems,
which include a complex floral element (Fig. 18a, b, c, d).
With the ability to identify these marsh flora elements in
present surface morphology of the tidal flat marshes and in
the subsurface, plus a precise understanding of internal structure of adjacent barriers and nearshore marine and estuarine
sediments, we can now begin to create detailed paleogeographic
sedimentary and environmental lithosome distributions throughout the entire group of models that are developed in the Delaware coastal zone.

The technique for microanalysis of salt marsh flora is
described in detail by Allen (1977).   Figure 19 shows examples
of three marsh floral elements.   Microsections of living stem
specimens of *Spartina alterniflora, Spartina patens, and
Phragmites communis,* are compared with microsections of paraffin
embedded marsh core specimens showing the equivalent specimens
in a partially decayed situation.   The examples given in Figure
19 are, of course, ideal examples.   However experience has
shown that a large majority of the floral elements of 17 salt
marsh floral facies and sub-facies can be identified from organic debris and marsh sediments.   In washover barrier-broad

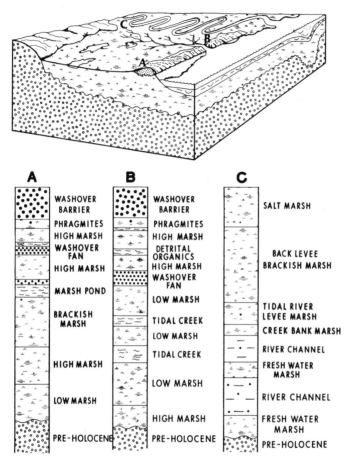

Fig. 17. TYPICAL VERTICAL SEQUENCES OF MARSH ENVIRONMENTS.
A block diagram interpretation showing vertical sequences of
marsh environments related to shoreline types along Delaware
Bay. Sequence A represents marsh development landward of a
continuous barrier shoreline. Sequence B is located along a
discontinuous barrier shoreline and Sequence C developed along
a tidal river-marsh shoreline.

marsh sequences (Fig. 17, A and B), high and low-marsh environments are the most important sediment producers. Brackish-
marsh environments may contribute a significant amount to the
sediments, but fresh-water marsh sediments are a minor component. In the tidal-river sequence (Fig. 17, C), brackish-
marsh environments are the most important sediment producers.

(a)

(b)

*Fig. 18. (a) A brackish water pond at the landward edge
of a broad salt marsh. The area is comprised of a highly complex and varied flora. (b) A high salt marsh meadow typical
of the* Distichlis spicata *and* Spartina patens *marshes of the
Delaware Estuary. (c) A low salt marsh of* Spartina alterniflora *along the edge of a tidal creek. (d)* Phragmites communis *marsh developing on the back of a very thin and narrow estuarine washover barrier. The root system of the* Phragmites
communis *marsh is exposed in the eroding tidal flat area in
front of the thin ephemeral washover barrier.*

Fresh-water marshes and high-low marshes probably contribute
significant and relatively equal amounts to the sediment. Stratigraphy of these organic coastal deposits indicates that it is
extremely important to know the characteristics of basal marsh
deposits if the material is to be radiocarbon dated and used
for construction of sea-level curves. For instance, some basal
organic-rich sediments had the fibrous texture of *Spartina
alterniflora*; yet in microtome sections the same sediments consisted entirely of ferns and fern debris. The time framework

(c)

(d)

Fig. 18c,d

between deposition of the fresh-water ferns and the overlying estuarine sediments may only be a few hundred years; yet, it also could possibly be on the order of a thousand years and have very little to do with sea-level change.

Estuarine Marsh-Cliff Shoreline

The northernmost model of the Delaware coast discussed in this paper is that of an estuarine marsh and tidal flat at the shoreline of the upper portion of Delaware Bay and the lower tidal Delaware River. In this area, broad intertidal flats with relict marsh surfaces are exposed. Occasionally narrow "pocket beaches" occur where a local sediment source exists (Fig. 20). The area consists of a broad coastal marsh adapted to tidal drainage systems and the low-lying topography of the

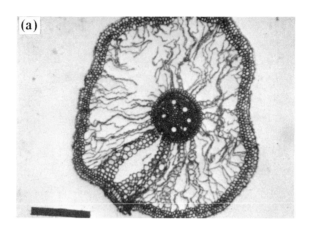

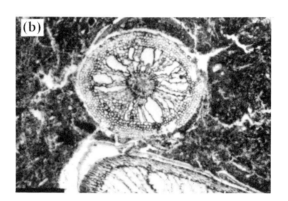

Fig. 19. Examples of microsections of salt marsh floras:
(a) Spartina alterniflora, *a modern root; (b) a well-preserved*
Spartina alterniflora *root in marsh sediment; (c)   a microsec-*
tion of Spartina patens, *stem of a living plant; (d) a slightly*
dessicated stem of Spartina patens *in salt marsh sediments; (e)*
a cross-section of a portion of the stem of living Phragmites
communis; *(f) a microsection of a portion of the stem of*
Phragmites communis *from a subsurface marsh sediment sample.*
Scale bar on all photographs = 400 microns.

mid-Delaware coastal plain (Fig. 1). Floral studies of marsh
facies show broad areas of high and low tidal marsh, and brackish marsh, as one approaches the leading edge of the transgression. A dense *Phragmites* marsh is developed along the shore
zone and in those areas where narrow sandy washover barriers

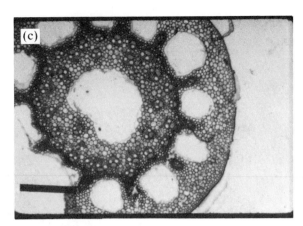

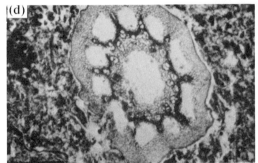

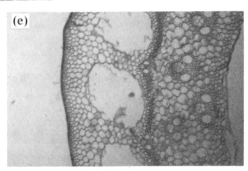

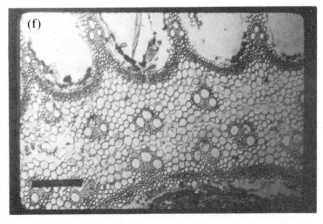

Fig. 19c,d,e,f

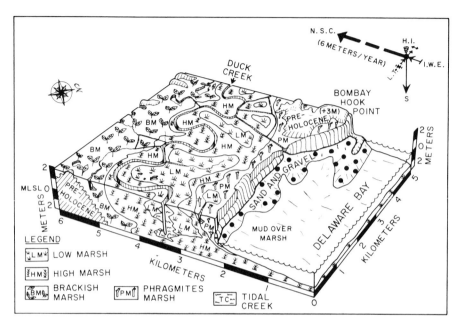

Fig. 20. ESTUARINE MARSH-CLIFF SHORELINE. A block diagram of a broad coastal marsh of the upper Delaware Bay and lower Delaware tidal river in an area of low coarse sediment supply source. Elements of the varied living salt marsh flora are shown in relationship to a subsurface interpretation of marsh facies. The Bombay Hook point is a local source of coarse size sediments. Coarse sediment supply is not enough to form a continuous barrier. Broad tidal flats of eroding marsh sediments occur in front of the barrier. The vector diagram shows the relative importance of various processes in action in this type of coastal setting. Compare the vector diagram with Figure 3 for key to the symbols.

occur. These facies can be traced by drilling into the sub-surface (Fig. 20). The tidal range is slightly higher than in lower Delaware Bay but there is a relative decrease in wave ac-tion as one approaches the narrower tidal Delaware River. Fur-ther, waves tend to break bayward on the broad tidal flats, de-pending on the tidal position. A small amount of sand and gra-vel is derived from local highlands such as at Bombay Hook Point. The amount of sand and gravel available for barrier construction is generally insufficient to maintain a continuous barrier.

   Large volumes of silt and clay from the salt marsh are added to the turbid waters of Delaware Bay and recycled into the tidal marshes through tidal creeks and by flooding at spring

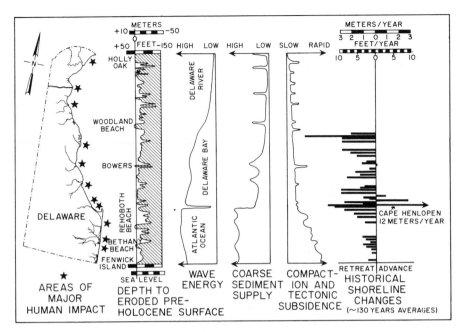

Fig. 21. A composite diagram relating the various coastal settings of Delaware to areas of major human impact, depth to the pre-transgression surface, wave energy, coarse sediment supply, compaction and tectonic subsidence, and historical analysis of shoreline changes.

tides.   Thus the tidal-marsh surface level tends to keep pace with the local relative rise of sea level as the transgression continues.   Human intrusion in this area is very low as it is not a desirable area for human habitation.   Incident wave energy is relatively low, but net shoreline change or coastal erosion averaging six meters per year occurs as the waves impinge directly on a muddy shoreline at high tide.

PROCESS AND FORM

    Analysis of dynamic processes; sediments available for erosion, transportation and deposition; initial topographies undergoing transgression and resultant morphologies; and subsurface studies allow for a determination of variable elements of evolution of the different coastal settings of the Delaware coastal zone.   Sand barriers occur as a function of availability of sediment and wave and current energy.   Shorelines lack

coarse sediment in some areas such as the Delaware River estuary, where marshes form the shoreline edge. Highly varied
models have been developed to delineate the varied geomorphic
and geologic configurations of the Delaware coastal zone. Figure 21 presents a generalized summary diagram of some of the
observed elements of coastal development. Areas of major human impact are essentially those areas in which man has attempted to develop small coastal villages or cities or where man
has placed groins, jetties, or breakwaters. The configuration
of the topography undergoing transgression is extremely important in the development of the transgressive coastal environmental lithosomes. Depths to eroded pre-Holocene surface, as
shown on Figure 21, show the axes of tributary valleys to the
ancestral Delaware River as they cross the Delaware coastal
plain toward the center of present Delaware Bay, where the axis
of the late Wisconsin ancestral Delaware River was positioned.
From this curve showing depth to eroded pre-Holocene surface,
it can be observed that there are a small number of highland
surfaces undergoing erosion in the beach or shoreface area. On
the other hand, the valley systems undergoing transgression
provide loci for accumulation of relatively thick marsh muds.
In the lower part of Delaware Bay, some of these wider valleys
were at one time open water bodies or coastal lagoons before
being filled by tidal salt marsh (Elliott 1972). It is possible that the closing of the lower Delaware Bay lagoons, which
were present as late as 300 to 400 years ago, may have been
accelerated by the intrusion of aboriginal American or European
settler's agricultural activity.

As expected, wave energy is high along the exposed Atlantic
Ocean coast of Delaware. Wave energy levels sharply decrease
in the protected area of Lewes Harbor. A gradual decrease in
incident wave energy occurs as one proceeds northward into Delaware Bay as the bay narrows toward the tidal Delaware River.

Volume of coarse sediment supply is extremely important in
the development of barriers. Major barriers are developed along
the Delaware Atlantic coast as shown in the coarse sediment supply graph on Figure 21. Coarse sediment supply peaks shown on
the curve in Figure 21 correspond to areas in which necks of
land or highland portions of a coastal plain crop out along the
shoreline.

Compaction and tectonic subsidence is believed to have been
relatively uniform across the Delaware coastal plain (Belknap
and Kraft, 1977). As one approaches the Atlantic coast and the
nearshore marine offshore area, there is evidence of a significant increase in rate of subsidence. This may be due to tilting
of the continental shelf because of water loading, or because of
tectonic factors which might include coastal parallel faulting
(Belknap and Kraft, 1977).

Historical shoreline changes over the past 130 years can be

well documented, based on records of the U.S. Coastal Survey, the U.S. Coast and Geodetic Survey, the National Ocean Survey, and the U.S. Army Corps of Engineers (1956). As seen in Figure 21, the shoreline is one of dominant coastal erosion and transgression. In some areas along Delaware Bay, shoreline retreat rates reach extremes of up to six meters per year; but the long-term average of three meters per year occurs along most of the Bay shoreline. Along the Atlantic coast, erosion rates up to three meters per year occur in the Cape Henlopen area, and decrease as one proceeds southward along the coast to one to two meters per year over the long term average. Local accretion, such as in the case of the Cape Henlopen spit, averaged twelve meters per year over the past 200 years, but is known to advance up to 30 meters per year at present. Other small points of local accretion along the Delaware Bay coast are due to man's intrusion by beach nourishment or the development of groins.

CONCLUSIONS

A genetic classification which includes a number of models has been developed for transgressive coastal systems in the Delaware coastal zone. Because of complex interrelationships of many factors, highly varied models of evolution of the coastal transgression from over 100 kilometers offshore to the present position, dominant elements observed in the models developed appear to be relative sea-level change, incident wave energy, sediment supply, and the nature of the pre-Holocene surface undergoing transgression. Projections into the short-term geological future suggest that the coastal sedimentary lithosomes will continue to migrate landward until a point is reached where the lower portion of the Delmarva Peninsula is completely submerged. Studies of the pre-Holocene sediments suggest that a similar situation existed during the Sangamon Age.

REFERENCES

Allen, E.A. (1974). Identification of *Gramineae* fragments in salt-marsh peats and their use in late Holocene paleoenvironmental reconstructions. Unpublished M.S. thesis. Dept. of Geology, Univ. of Delaware, Newark, 141 p.

Allen, E.A. (1977). Petrology and stratigraphy of Holocene coastal marsh deposits along the western shore of Delaware Bay. Delaware Sea Grant Tech. Rept. DEL-SG-20-77. College of Marine Studies, Univ. of Delaware, Newark, 287 p.

Belknap, D.F. and Kraft, J.C. (1977). Holocene relative sea-level changes and coastal stratigraphic units on the north-west flank of the Baltimore Canyon trough geosyncline. *J. Sed. Petrology 47*, 610-629.

Demarest, J.M. (1978). The shoaling of Breakwater Harbor-Cape Henlopen area, Delaware Bay, 1842 to 1971. DEL-SG-1-78. College of Marine Studies, Univ. of Delaware, Newark, 169 p.

Dolan, R. (1973). Barrier islands: Natural and controlled. *In* "Coastal Geomorphology"(D.R. Coates, ed.), p. 263-278. Pub. in Geomorphology, State Univ. of N.Y., Binghamton.

Elliott, G.K. (1972). The Great Marsh, Lewes, Delaware. Tech. Rept. No. 19. College of Marine Studies, Univ. of Delaware, Newark, 139 p.

Fischer, A.G. (1961). Stratigraphic record of transgressing seas in light of sedimentation on Atlantic coast of New Jersey. *Amer. Assoc. Petroleum Geologists Bull. 45,* p. 1656-1666.

Fisher, J.J. (1967). Origin of barrier island chain shorelines, Middle Atlantic states (abs.). Geol. Soc. America, Spec. Paper 115, p. 66-67.

John, C.J. (1977). Internal sedimentary structures, vertical stratigraphic sequences, and grain-size parameter varia-tions in a transgressive coastal barrier complex - The Atlantic Coast of Delaware. DEL-SG-10-77. College of Marine Studies, Univ. of Delaware, Newark, 287 p.

Kraft, J.C. (1971). Sedimentary facies patterns and geologic history of a Holocene marine transgression. *Geol. Soc. America Bull. 82*, p. 2131-2158.

Kraft, J.C. (1976). Radiocarbon dates in the Delaware coastal zone (eastern Atlantic coast of North America). DEL-SG-19-76. College of Marine Studies, Univ. of Delaware, 20 p.

Kraft, J.C. and Caulk, R.L. (1972). The evolution of Lewes Har-bor. Tech. Rept. No. 10. College of Marine Studies, Univ. of Delaware, 58 p.

Kraft, J.C., Biggs, R.B., and Halsey, S.D. (1973). Morphology and vertical sedimentary sequence models in Holocene trans-gressive barrier systems. *In* "Coastal Geomorphology" (D.R. Coates, ed.), p. 321-354. State Univ. of N.Y., Binghamton.

Kraft, J.C. and Belknap, D.F.(1975). Transgressive and regres-sive sedimentary lithosomes at the edge of a late Holocene marine transgression. Extraits des Publications du Congres, IX Congres International de Sedimentologie, Nice, p. 87-95.

Kraft, J.C. and John, C.J. (1976). The geological structure of the shorelines of Delaware. DEL-SG-14-76. College of Marine Studies, Univ. of Delaware, Newark, 106 p.

Kraft, J.C., Allen, E.A., Belknap, D.F., John, C.J., and Maurmeyer, E.M. (1976). Delaware's Changing Shoreline. Tech.
Rept. No. 1. Delaware's Coastal Management Program, Delaware State Planning Office, Executive Dept., Dover, 319 p.

Kraft, J.C., Allen, E.A., and Maurmeyer, E.M. (1978). The geological and paleogeomorphological evolution of a spit system and its associated coastal environments, Cape Henlopen
Spit, Coastal Delaware. *J. Sed. Petrology 48*, 211-226.

Maurmeyer, E.M. (1974). Analysis of short and long term elements
of coastal change in a simple spit system: Cape Henlopen,
Delaware. Unpubl. M.S. thesis. Dept. of Geology, Univ. of
Delaware, Newark, 149 p.

Turner, P.A. (1968). Shoreline history, Atlantic coast, Delmarva Peninsula (abs.). 1968 Annual Meeting, NE Section,
Soc. Econ. Paleon. and Mineralogists, Washington, D.C.

U.S. Army Corps of Engineers (1956). Delaware Coast from Kitts
Hummock to Fenwick Island, beach erosion control study.
U.S. Army Corps of Engineers, Philadelphia, Pennsylvania,
47 p., Appendices A-H.

NEXUS: NEW MODEL OF BARRIER ISLAND DEVELOPMENT

Susan D. Halsey[1]

Department of Geology
University of Delaware
Newark, Delaware

Geomorphic and stratigraphic data from the oceanic coast of the Delmarva Peninsula show that the configuration of the pre-Holocene erosion surface has had a strong influence on the type of transgressive barriers that are developing along the coast. This linking, hence nexus, of the new topography with the old, combined with a differing sediment supply along the length of this coast, helps to explain the similarity of landforms along the compartments of the mid-Atlantic Bight.

Coring reveals a relict topography of mid-Wisconsinan(?) age, whose transgressive phase left environments and lithosomes similar to those in the Holocene. The regressive phase (late Wisconsinan) left a variable density paleochannel network with higher interfluve divide areas. As the Flandrian-Holocene relative transgression began, beaches formed against highlands of the divides, and estuaries developed in the paleochannel valleys. Major inlets remained in the relict thalwegs until sediment supply and coastal dynamics were sufficient to produce closure. In the northern section of the Peninsula the relict topography is characterized by deeply incised headland areas with a relatively dense dendritic drainage pattern. As a result, the Holocene pattern of landforms consists of beaches against highlands, a northern spit, small lagoons, and baymouth barriers linking headlands. The central section, along Assateague Island, had a coarser relict drainage pattern, resulting in the earlier-late Holocene barriers being segmented and separated from the mainland by a wide lagoon. Relative sea level rise, continued sediment supply and spit growth allowed these

---

[1]*Present address: New Jersey Department of Environmental Protection, Bureau of Geology & Topography, Trenton.*

segments to link. The relict topography of the southern barrier island chain was a wave cut cliff-shoreline with paleochannels incised perpendicular to the coast. Because of a decrease in sediment supply and lack of significant headlands, the segments of the barrier island chain below Assateague Island have remained unlinked throughout the Holocene transgression and the high number of inlets has allowed vast marsh and tidal flat development to occur landward of the barrier.

INTRODUCTION

The Nexus model of barrier island development is an outgrowth of work on previous theories of barrier island origin as compiled in Schwartz (1973) and those of more recent publication (Swift 1975; Field and Duane, 1976). These discussions do not fully reflect either the transgressive, migrational, and dynamic nature of the barrier systems along our coasts or the influence the configuration of the pre-Holocene erosion surface has on the new barriers of the Holocene. Although Hoyt's (1967) submergence model of barrier island formation was considered incomplete, Fisher's (1967) spit building mechanism did not provide a satisfactory origin either. One element of Fisher's (1967) model, his coastal compartment model (Fig. 1), was very interesting. That model suggested the various landforms of the compartments of the mid-Atlantic Bight were genetically related. The Nexus model is an attempt to explain the genetic reasons for the barrier systems of the Delmarva Peninsula and then to suggest the reasons for the obvious similarity of the coastal landforms in the other compartments of the mid-Atlantic Bight as well as other barrier systems on passive continental margins (Inman and Nordstrom, 1971).

METHODS

Stratigraphic cross-sections using 135 cores were prepared and interpreted perpendicular to the shore, and parallel to the coast of the Delmarva Peninsula from Cape Henlopen, Delaware to near Cape Charles, Virginia (Fig. 2). In addition to the cores and sections, aerial photographs, charts, maps, and historical data were used to determine the geologic history and morphologic development of the latest pre-Holocene and Holocene environments of coastal Delmarva (Halsey 1978). The transgressive vertical sedimentary sequence models of barrier systems as presented by Kraft and others (1973) were used to interpret the cross-sections.

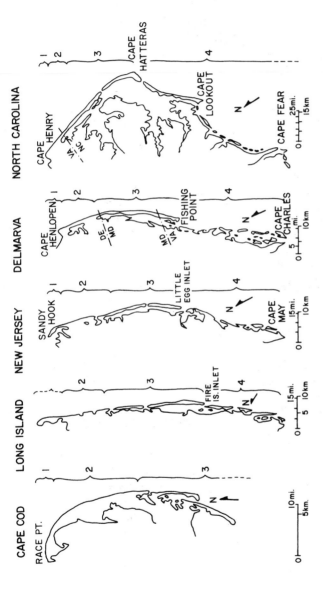

COASTAL COMPARTMENTS OF THE MIDDLE ATLANTIC BIGHT

1: NORTHERN OR CUSPATE SPIT   3: SOUTHERN "SPIT"

2: ERODING HEADLAND   4: BARRIER ISLAND CHAIN

modified from FISHER, J.J., in SWIFT (1969)

Fig. 1. Modified Fisher (1969) diagram delineating the geomorphic similarity of the coastal compartments of the Mid-Atlantic Bight.

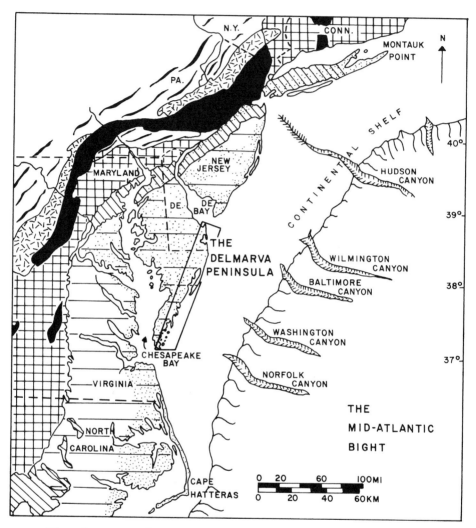

Fig. 2. Location map of the study area in the Mid-Atlantic Bight. Key: Dots=Quaternary; Horizontal lines=Tertiary: Eocene and Miocene; Diagonal lines=Upper and lower Cretaceous; Black= Triassic and Jurassic; Wavy=Paleozoic; Boxes=Paleozoic-Precambrian; Random=Precambrian. Modifed from Kraft (1971b).

INTERPRETATION OF DATA

It was determined from interpretations of the cross-sections that the configuration of the pre-Holocene erosion surface along the Delmarva Peninsula had a strong influence on the type of transgressive barriers that developed along this compartment as well as other compartments along the East Coast (Halsey 1978). A linking or connection was seen to occur between the pattern of the regressive late Wisconsinan drainage patterns with intervening interfluve divide areas and the types of Holocene barrier systems that developed on, over, and around the relict topography. This linking of the old topography with the new topographic features is the basis of the name of the model: nexus, which means "1: connection, link, or 2: a connected group or series" (Webster 1963). In addition to the general but primary connection of the topographies, a second link on a more local scale can be seen between the pre-Holocene environments and the dynamics of the latest barrier island systems, especially sediment supply. A third link can be postulated for the future. If sea level continues to rise, the Holocene transgressive barriers will approach and intersect the dissected remnants of a substantially greater portion of the higher pre-Holocene coastal and highland environments. Thus, there appears to be a progressive evolution of barrier forms initially influenced by pre-Holocene topography.

Pre-Holocene Paleogeography and Sedimentation

The latest pre-Holocene sedimentary environmental lithosomes of the Delmarva Peninsula have two components, one a transgressive phase, and the other a regressive phase. The cycle was proposed to be of mid-Wisconsinan (Silver Bluff) -late Wisconsinan age based on correlation with the revised provisional stratigraphic sequence for the lower Delmarva Peninsula by Owens and Denny (1974), and Table I. In addition, radiocarbon dates of organic-rich sediments along the length of Delmarva clustered closely on the Hoyt and Hails (1974) revised Silver Bluff submergence curve (Fig. 3). Other data to support this age determination were type of fauna, degree of compaction, and correlation to other geographic localities (Halsey and others, 1977; Halsey 1978; Rampino 1973; Rampino and Sanders, 1976, 1977; Moslow and Heron, 1977, 1978, this volume; Susman and Heron, 1979).

The mid-Wisconsinan transgressive phase is characterized by coastal environments and geomorphic features similar to those found in the Holocene. Silver Bluff environments encountered by the drill are lagoonal-marsh lithosomes, possible tidal delta-shoal, and some sands and gravels. Remnants of what is interpreted to be the highest stand of the mid-Wisconsinan barrier island system can also be found. However, most of the mid-Wisconsinan barrier sands cannot be found in drill holes through the present Holocene barriers because of the higher

TABLE I. Revised provisional stratigraphic sequence in the Maryland-Delaware parts of the lower Delmarva Peninsula (by Owens and Denny, 1974).

			KEY: (fe): fluviatile-estuarine (bb): barrier-back barrier (oo): open ocean-marine
PLEISTOCENE-HOLOCENE	LATE WISCONSINAN – HOLOCENE	CHESAPEAKE AND DELAWARE BAY FILL (fe) PRESENT COASTAL SYSTEM (bb) FILL OF LARGER COASTAL RIVERS (fe)	
	MID-WISCONSINAN (?)	BURIED BARRIER SYSTEM UNDERLYING PART OF ASSATEAGUE ISLAND	
		PEATS AND SANDS OF THE SINEPUXENT NECK (bb,oo)	
	SANGAMONIAN	II. SANDY DEPOSITS IN RIVER VALLEYS AND SILTY AND SANDY DEPOSITS IN FRINGING BAYS, TOP ABOUT 6 M ABOVE SEA LEVEL (fe) I. SANDY DEPOSITS (BEACH AND/OR DUNE) FRINGING LOWER PENINSULA, TOP ABOUT 12 M ABOVE SEALEVEL (fe) OMAR FORMATION OF JORDAN,1962 (bb)	
PLIOCENE		WALSTON SILT (bb)	
		BEAVERDAM SAND (bb)	
		POKOMOKE AQUIFER (oo)	
MIOCENE		PENSAUKEN (?) FORMATION (fe)	
		CHESAPEAKE GROUP UNDIFFERENTIATED (INCLUDES ALL AQUIFERS NAMED BY CUSHING AND KANTROWITZ,1973, EXCEPT POKOMOKE (oo)	

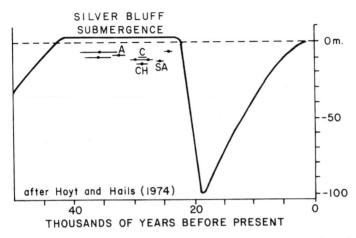

Fig. 3. Revised middle to late Pleistocene and Holocene sea level curve by Hoyt and Hails (1974) with data from Halsey (1978). Horizontal bars through dots represent ± deviation in years. A=Radiocarbon date ("peat") from northern Assateague Island, Maryland (core SDH-4-71); C=Radiocarbon date (wood) from Chincoteague Island, Virginia (core SDH-33-72); SA=Radiocarbon date (wood fragments, combined) from southern Assateague Island (core MSG-6-76).

level of the mid-Wisconsinan sea. Rather, the mid-Wisconsinan barrier sands can be found landward and higher in elevation than the presently seaward and lower barriers of the Holocene.

When paleogeographic maps are reconstructed for the Delmarva Peninsula during the highest stand of the mid-Wisconsinan sea level, at least three large barrier island complexes are suggested in the northern half of the Peninsula. They are named the "Rehoboth Barrier," the "Bethany Barrier," and the "Sinepuxent Barrier" (Fig. 4). They appear to have had a "drumstick"-like shape, with bulbous seawardly offset accreting northern ends, eroding central sections, and accreting spit and beach ridge growth on their southern ends, similar to Holocene barriers described by Hayes and Ruby (1975) and Hayes (this volume). These barrier segments were separated from each other by large estuarine-lagoon systems. The Indian River Estuary separated the Rehoboth and Bethany Barriers, and the St. Martin Estuary-Isle of Wight Bay system separated the Bethany and Sinepuxent Barriers. In the southern half of the Peninsula, in the area of the present barrier island chain from Wallops Island to Cape Charles, a wave cut cliff-shoreline, named the

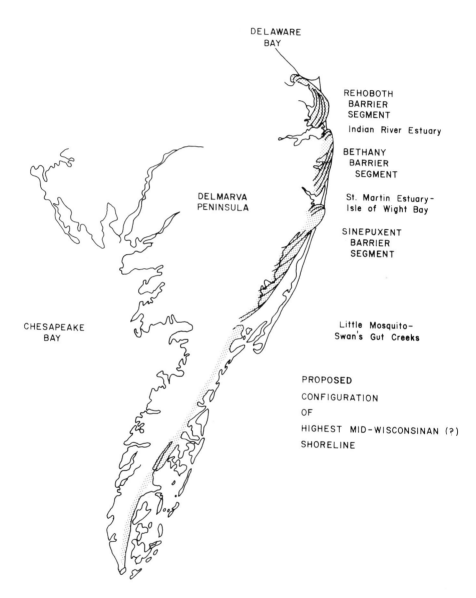

*Fig. 4. Proposed configuration of the highest stand of
the mid-Wisconsinan barrier system. Gray=barrier segments.
Arrows point to intervening estuary-bay areas between segments.*

"Virginian Shoreline-Barrier," was found with a large spit complex extending south of the present town of Wachapreague, Virginia (Halsey and others, 1977). The Little Mosquito Creek-
Swan's Gut Creeks area, just north of Wallops Station, Virginia
is thought to have been the estuarine-lagoon system separating
the southern extremity of the Sinepuxent Barrier from the Virginian Shoreline-Barrier.

The late Wisconsinan regression that followed the mid-
Wisconsinan high sea level stand began approximately 21,000 yrs.
BP (Hoyt and Hails, 1974) (Fig. 3) and was responsible for determining the course of migration and type of barriers in the
Flandrian-Holocene transgression. As a result of the regression,
much of the sandy portion of the mid-Wisconsinan barrier systems were eroded as estuarine, and then fluvial systems were
developed on the emerging continental shelf. Aggradation to the
mid-Wisconsinan barrier segments seems restricted to the development of the Bell-Bradford and Upshur Neck spit complex
along the Virginian Shoreline-Barrier, minor modification of
offset inlets along the Shoreline-Barrier, and possible regressive beach ridge deposits seaward of the toe of the Shoreline-
Barrier.

This is in contrast to Susman and Heron's (1976, 1979) and
Moslow and Heron's (1977, 1978, and this volume) interpretation
for coastal North Carolina. There is no evidence along the Delmarva Peninsula for lagoonal deposits (Diamond City) as suggested by Moslow and Heron (1978, and this volume) being deposited
during this time. Instead, the extensive lagoonal lithosome of
the Delmarva Peninsula, the blue-gray Chincoteague Island silt
which is both stratigraphically and lithologically correlative
with the Diamond City, was interpreted by Halsey (1978) as
transgressive mid-Wisconsinan, since the Hoyt and Hails (1974)
sea level curve indicates a rapid regression (Fig. 3).

Along the Delmarva Peninsula, this rapid regression left a
variable density paleochannel network with intervening higher
interfluve divide areas. In the northern section of the Peninsula from Cape Henlopen, Delaware to Ocean City, Maryland, the
pre-Holocene topography was characterized by deeply incised
headland areas with a relatively dense dendritic drainage pattern. This pattern may have been determined by the proximity
to the large ancestral Delaware River course, and can be seen
by examining the configuration of the pre-Holocene erosional
drainage surface from Cape Henlopen to Bethany Beach, Delaware
as determined from seismic studies by Sheridan and others
(1974) (Fig. 5). The paleochannels in the Cape Henlopen area
(Wolfe and Holland Glades, and Dupont Basin) as well as the
Love-Herring Creek, Indian River, and Salt Pond paleochannels
have been found by drilling under the present barrier system
(Halsey 1978). The channels' more landward courses are postulated

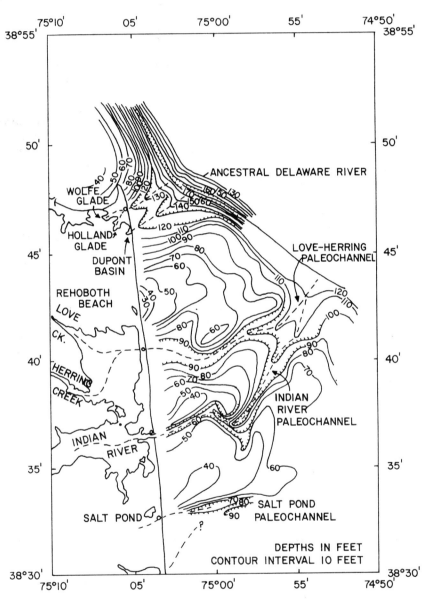

Fig. 5. *Configuration of the pre-Holocene erosional drainage surface as determined by seismic studies by Sheridan and others (1974) with paleochannel courses dashed in. Dots locate cores that found thalwegs of paleochannels.*

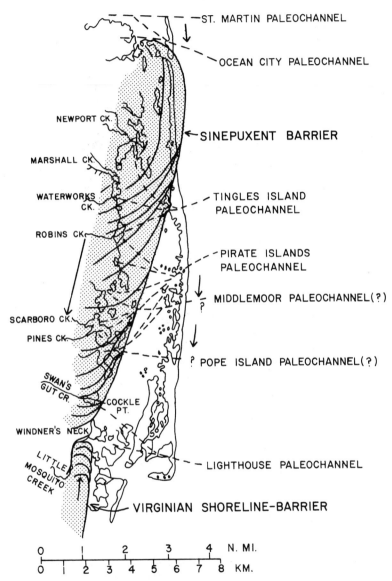

Fig. 6. *Proposed configuration (dots) and growth (heavy curved lines) of the Sinepuxent Barrier during the highest stand of the mid-Wisconsinan sea and the proposed drainage network (dashed lines) and possible migration (large arrows) of the paleochannels formed during the late Wisconsinan regression.*

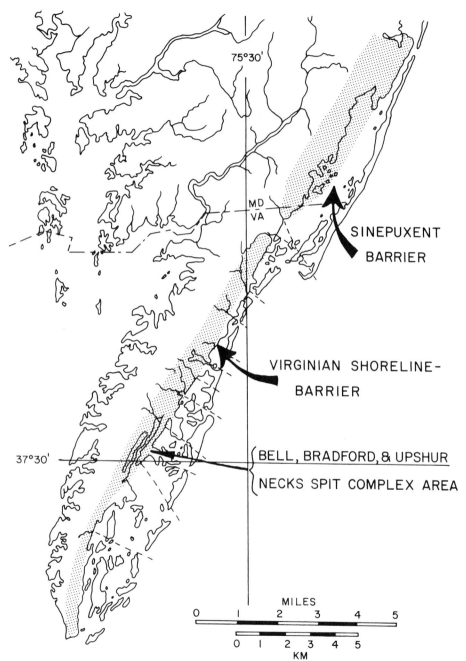

Fig. 7. Proposed configuration (dots) of Sinepuxent Barrier &
Virginian Shoreline-Barrier (highest stand of mid-Wisconsinan
sea & initial stages of late Wisconsinan regression). Dashed
lines=proposed paleochannel locations (late Wisc. regression)

by the underfit nature of the present creeks or rivers, and
their alignments. In the case of Indian River, a radiocarbon
date of 10,800± 300 yrs. BP on "peat" at the base of the Holocene section suggests that the Indian River paleochannel was
incised during the late Wisconsinan regression and was reoccupied during the Flandrian-Holocene transgression (Halsey 1978).

In the central section of the Peninsula from Ocean City
south along Assateague Island, the pre-Holocene topography appears to have had a coarser dendritic drainage pattern as compared to the more northern area (Fig. 6). This was probably due
to the fact that both the remnant headlands south of Sinepuxent
Neck and the mid-Wisconsinan barrier system were further landward near the western shore of present Chincoteague Bay. Also,
the main thalweg of the incising Delaware River was further removed from this more southerly location during the late Wisconsinan regression. Figure 6 shows the proposed configuration
of the Sinepuxent Barrier during the highest stand of the Mid-
Wisconsinan sea and the proposed drainage network of the paleochannels during the late Wisconsinan regression.

The St. Martin and Ocean City paleochannels, the Tingles
Island and Pirate Islands paleochannels, and the Lighthouse
paleochannel have been located by drilling (Halsey 1978). However, no drilling data is yet available for the Middlemoor and
Pope Island paleochannels, and their locations are proposed on
the basis of the configuration of the underfit creeks on the
western side of Chincoteague Bay. Furthermore, the southward
trending arrows (Fig. 6) and the overlapping nature of the
landward courses of the drainage patterns were drawn to suggest
a possible southward migration of the active channels during
the initial stages of the late Wisconsinan regression.

Because of the configuration of the wave cut Virginian
Shoreline-Barrier to the south, paleochannel incisement was
perpendicular to the coast, giving the upper reaches of the
streams a parallel-type drainage network with relatively flat
interfluve divides (Fig. 7). The proposed paleochannel locations seaward of the presently underfit streams emanating from
the Virginian Shoreline-Barrier are, for the most part, aligned
with the present inlet system. In addition, there is geomorphic
and stratigraphic evidence to support the existence of a very
late mid-Wisconsinan-earliest late Wisconsinan spit complex in
the Bell, Bradford, and Upshur Neck area that caused some
paleochannel reorientation in that area. However, it is not
known into which pattern these streams coalesced on the emerging inner shelf. It may be that the incising streams flowing
from the Virginian Shoreline area were not part of the late
Wisconsinan Delaware River system but were instead part of the
"Chesapeake" watershed.

Therefore, geomorphic and stratigraphic evidence reveal a

variable paleochannel-interfluve divide pattern along the coast
of the Delmarva Peninsula left as a result of the late Wisconsinan regression. This topography was present as the rising
waters of the Flandrian-Holocene transgression began to flood
the Delmarva coast.

Holocene Sedimentary Patterns

The Assateague Island-Chincoteague Bay area (Fig. 8) illustrates well the Nexus model because of its intermediate location between the northern deeply incised headland areas and the
southern more coarsely incised area. The three large dots seen
on the left of Figure 8 are in the same position on the succeed-
ing paleographic maps (Figs. 9-12).

The first step in the development of the Nexus model is
proposed to have begun when the rising waters of the Flandrian-
Holocene sea intersected the remnant cuestas and paleochannel
valleys produced during the late Wisconsinan regression (Fig. 9;
Table II). Using recent analogs from highland localities such
as Cape Cod and Martha's Vineyard, Massachusetts; eastern Long
Island; and Rehoboth Beach and Bethany Beach, Delaware (Fig. 1),
it can logically be assumed that beaches formed against the
lower slopes of the cuesta highlands of the interfluve divides,
and that narrow estuaries developed in the flooded paleochannel
valleys.

Sediment eroded from the headlands contributed to the longi-
tudinal growth of the developing barrier segments by spit accretion and ebb-tidal delta growth. Landward growth or migration
of the barriers was augmented by the contribution of sediment
to beaches, dunes, washovers, and flood-tidal deltas (Fig. 10).
As sea level continued to rise, tidal flats and lagoons developed in lower landward areas as the barrier islands slowly migrated landward, eroding the interfluve divides.

While the Nexus model recognizes inevitable submergence in
a transgression, Hoyt (1967) had proposed that it was necessary
for a newly formed barrier island to have a 6 meter high dune
field in order to survive submergence and landward migration.
However, more recent studies on migrating barrier islands, especially work by Kraft (1971a, 1971b), Kraft and others (1973),
Godfrey and Godfrey (1973), Field and Duane (1976), Leatherman
(1976), as well as others, have shown that it is not necessary
to have a high dune field to perpetuate a barrier island as he
proposed. Rather, unaltered barrier islands in a relative transgression have been shown to migrate landward and upward in
space and time in an *in toto* fashion reflecting relative sea
level rise. The certain presence of eroding headlands acted as
cores to the proto-barrier islands, and those features appear to
be the sole requirement to form the first line of barriers.

Fig. 8. REFERENCE MAP FOR PALEOGEOGRAPHIC MAPS. Reference map for the paleogeographic maps of the Chincoteague Bay-Assateague Island area showing the present configuration of coastal landforms in latest Holocene times. Large reference dots are located in the same position on all paleogeographic maps.

It is proposed that as the barriers migrated landward over the interfluve divides, estuaries maintained their positions in the relict paleochannel valleys (Fig. 11). Therefore, the major inlets between barrier segments remained relatively fixed until such time as sediment supply was sufficient and coastal dynamics favorable enough to permit closure of the valleys by the action of tidal delta, spit platform, and spit accretion (Table II).

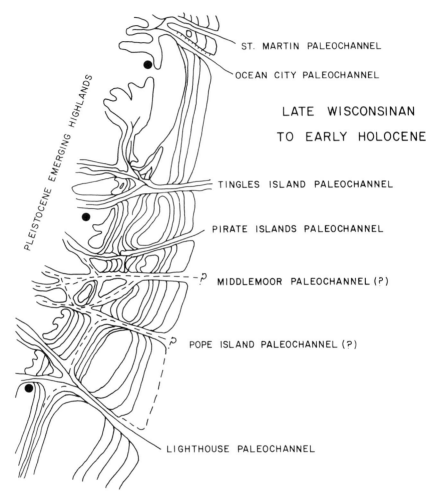

Fig. 9. Proposed paleotopography and paleochannel locations in the Chincoteague Bay-Assateague Island area during late Wisconsinan to early Holocene times. Large black dots for reference location, see Figure 8.

DISCUSSION

The amount of sediment supply, garnered mainly from the erosion of pre-Holocene highlands, seems to be the controlling factor in linking the old topography to the configuration of the new features. In the northern section characterized by

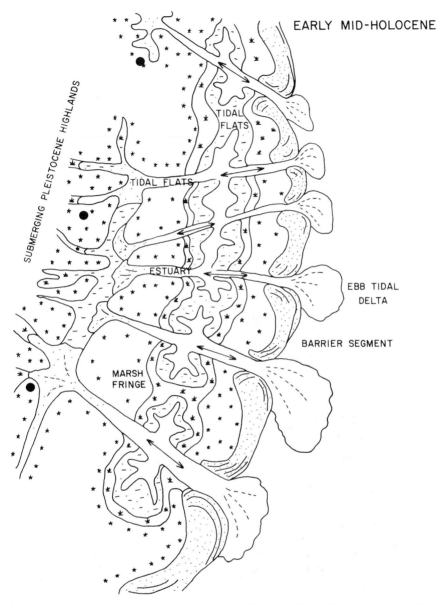

Fig. 10. Proposed paleogeography of the Chincoteague Bay-Assateague Island area during transgressive early mid-Holocene times. As sea level rose, former paleochannel locations became narrow estuaries with tidal delta deposits; lower areas filled with tidal flat and marsh fringe deposits, and barrier segments formed against the interfluve divide areas. Arrows denote direction of sediment movement.

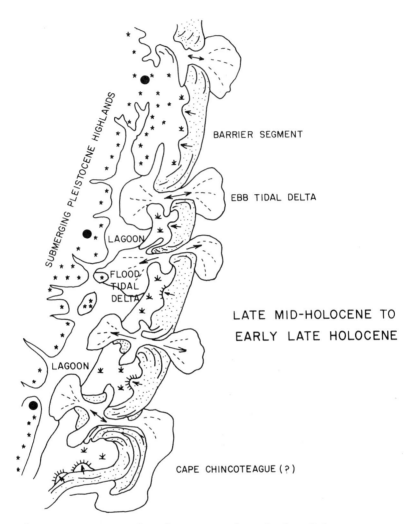

Fig. 11. Proposed paleogeography of the Chincoteague Bay-Assateague Island area during transgressive late mid-Holocene to early late-Holocene times. Note presence of proposed configuration of Cape Chincoteague. Arrows denote direction of sediment movement.

deeply incised headlands, and accordant large sediment supply, the Holocene coastal landforms that developed are: beaches against highlands, a northern spit complex, relatively small lagoons, and baymouth barriers linking headlands. In the central section along Assateague Island, where there was a coarser relict drainage pattern, it is proposed that the earlier late-

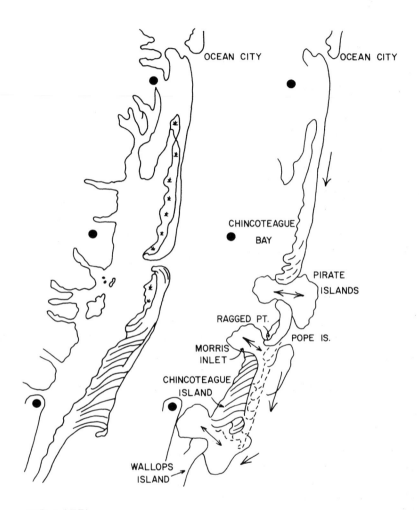

MID-LATE
HOLOCENE

EARLY COLONIAL TIMES

Fig. 12. Proposed configuration of the transgressive barriers in the Chincoteague Bay-Assateague Island area in mid-late Holocene (left) and early Colonial times (right). This linking would be an example of the second link in the Nexus model. Arrows indicate direction of sediment movement. Dashed lines indicate growth of the secondary barrier of Assateague Island seaward of the primary barrier of Chincoteague Island and the development of the Fishing Point spit complex.

TABLE II. Nexus Model of Barrier Island Development

---

*LINK I: Intersection of headlands and paleochannel valleys by
          rising sea produces...
          Beaches against highlands, narrow estuaries*

*LINK II: Continued sea level rise causes landward migration over
          interfluve divides, flooding of low areas, and mainte-
          nance of estuary positions, producing...
          Barrier islands, tidal deltas, estuaries and shallow
          lagoons, fringing marshes*
   A. **Northern area:** *deeply incised headlands, high inter-
      fluves produces...
      Beaches against highlands, baymouth barriers, small
      lagoons, northern spit complex*
   B. **Central area:** *coarser relict drainage pattern, lower
      and wider interfluves produces...
      Initially .segmented barriers with large lagoon
      landward.
      Continued sediment supply allows arcing of inlets
      and linking of barriers (nexus-spit)*
   C. **Southern area:** *backed by wave-cut cliff, low inter-
      fluves, perpendicular paleochannels, low coarse
      sediment supply produces...
      Segmented barrier chain with stable inlets, large
      number of inlets maintains large tidal prism,
      therefore large marsh and tidal flat development.*

*LINK III: Future: continued migration and intersection of higher
           topography produces increased sediment supply, filling
           of lagoons & estuaries, smoothened headland coastlines.*

Holocene barriers were segmented like the present barrier
island chain to the south but with a large lagoonal area behind. Subsequently, relative sea level rise and continued sediment supply seemed to have reached a critical coupling. This
situation has allowed most of the seaward ends of the former
paleochannel areas, acting as inlets, to close by continued
longshore drift and spit action. This action allowed the segments to fill or "arc" the former inlet areas and permitted
coalition of the segments into a long barrier island known as
a Nexus-spit (Halsey 1978). This is the present Assateague
Island (Fig. 12). Historical data and old maps indicate that
this may have happened in the Delmarva area as late as early
Colonial times (Truitt 1968, 1971; Wroton 1970). This would be
an example of the second link of the Nexus model (Table II).
   Continued inlet breaching, migration, and healing to the

south of the Pirate Islands-Green Run Inlet area allowed the
building of the extensive spit complex at the southern end of
Assateague Island, including Fishing Point, seaward of Chincoteague Island since the early 1800s (Gawne 1966, *in* Shepard
and Wanless, 1971; Kraft and others, 1973)(Fig. 12, right).
This rapid growth indicates the dynamic nature of coastal areas
when sediment supply is quite large.

In the southern area of the Peninsula along the barrier
island chain, the relict topography consisted of a higher wave-
cut cliff with lower interfluve divides seaward between paleochannels aligned normal to the coast (Fig. 7). A decrease in
sediment supply from the north and the lack of eroding headlands has resulted in the segments of the barrier island chain
remaining unconnected throughout the Holocene transgression.
Large tidal prisms due to the flattened interfluve divides and
the relatively large number of inlets of this chain have
allowed continued tidal flushing in spite of a large fine-
grained sediment influx. These conditions have resulted in the
extensive tidal flat and marsh development that are seen today
landward of the barrier island chain in contrast to the large
open water lagoon like Chincoteague Bay to the north (Fig. 12,
Table II).

Placement of the Nexus Model

In the overall scheme of barrier island origin theories,
the author would discourage the placement of the Nexus model in
Schwartz's (1971) multiple causality model under the "III. Composite" heading, even though it is a model of synthesis. This
is because of a fundamental difference both with the term and
processes implied in the "engulfed beach ridges" model as suggested by Hoyt (1967) as well as the "emergent bar" models.
Although the submergence of lower areas landward is recognized
and incorporated into the Nexus model, the configuration of the
underlying topography and sediment supply are considered more
critical factors. Since Hoyt's model was derived primarily from
the Georgia coast, which would be classified as a barrier
island chain, those aspects of the Nexus model applicable to
barrier island chains would be similar in some respects to
Hoyt's model.

Although the Nexus model includes spit growth as an important component, the model does not acknowledge Fisher's (1967)
contention that long prograded and ultimately breached spit action alone builds barriers. Instead, the linking of primary
barrier segments into what looks like a "spit," with subsequent
migration, breaching, and healing is favored by the Nexus
model. The concept of emergent offshore bars in a relative sea
level rise is not incorporated into the model because the

evidence is so meager (Otvos 1977; Field and Duane, 1977). However, due to the linking of the underlying topography to the
configuration of barriers, the Nexus model would probably apply
in coastal areas that may have undergone an earlier short-
lived Holocene relative sea level fall.

SUMMARY

Morphologic and stratigraphic evidence have shown that the
relict pre-Holocene topography has had a substantial effect on
the nature of the transgressive Holocene coastal landforms and
barrier systems, and will probably continue to do so as long
as there is a relative sea level rise. As Fisher (1967, 1968,
1977) indicated, the coastlines of the East and Gulf Coasts
have similar coastal landforms, as do most of the other "trailing edge"-Coastal Plain coasts. It is proposed that relict topography coupled with varying sediment supply has played a most
important role in the configuration and migration of Holocene
barrier systems.

Thus, Nexus is a new model that also combines aspects of
many researchers' work including Hoyt (1967), Biggs (1970),
Kraft (1971a, 1971b), Otvos (1970), Zeigler (1959), Fisher
(1967, 1968), Riggs and O'Connor (1975), Pierce and Colquhoun
(1970), Kaczorowski (1975) and Field and Duane (1976).
Although the Nexus model was developed to explain the configuration of the barrier systems along the Delmarva Peninsula, it
suggests that the Nexus model applies equally well to other
coastal compartments of the world. Indeed, preliminary data
indicate the model is operative in New Jersey (Halsey and
others, 1977, 1979) and in Long Island (McCormick, personal
communication).

ACKNOWLEDGMENTS

I wish to thank Dr. J.C. Kraft, my dissertation advisor at
the University of Delaware, and the other members of my com-
mittee, especially Dr. R.E. Sheridan, for their helpful comments during the initial stages of this work. The study was
funded by a grant to Dr. Kraft through Office of Naval Research.
Thanks also go to Dr. C. Larry McCormick and Dr. H. Allen Cur-
ran for their careful reviews of the manuscript, and to Mrs.
Julie Britt, Smith College, for typing.

REFERENCES

Biggs, R.B. (1970). The origin and geologic history of Assateague Island, Maryland and Virginia: Assateague Ecological
Studies, Final Report, Part I, Univ. of Maryland, Natural
Resources Institute, Cont. No. 446, p. 8-41.
Field, M.E. and Duane, D.B. (1976). Post-Pleistocene history of
the United States inner continental shelf: significance to
origin of barrier islands. *Geol. Soc. America Bull. 87*,
p. 691-702.
Field, M.E. and Duane, D.B. (1977). Post-Pleistocene history
of the United States inner continental shelf: significance
to origin of barrier islands. Reply. *Geol. Soc. America
Bull. 88*, p. 734-736.
Fisher, J.J. (1967). Origin of barrier island chain shorelines:
Middle Atlantic States. Geol. Soc. America Sp. Paper 115,
p. 66-67.
Fisher, J.J. (1968). Barrier island formation: Discussion.
Geol. Soc. America Bull. 79, p. 1421-1426.
Fisher, J.J. (1977). Atlantic Coast barrier islands as depo-
sitional models for oil exploration. Poster Session, 1977
Annual Mtg., Amer. Assoc. Petrol Geol., Washington, D.C.
Godfrey, P.J. and Godfrey, M.M. (1973). Comparison of ecologi-
cal and geomorphic interactions between altered and unal-
tered barrier island systems in North Carolina. *In* "Coas-
tal Geomorphology" (D.R. Coates, ed.), p. 239-258. Publ.
in Geomorphology, Binghamton, N.Y.
Halsey, S.D. (1976). Late Pleistocene and Holocene geologic
history and morphologic development of the Chincoteague-
Assateague area of Maryland-Virginia with implications to
the origin of barrier islands. *In* "Women in Geology"
(Halsey and others, eds.), p. 45-57. Proceedings of the
First Northeastern Women's Geoscientists Conference, Ash
Lad Press, Canton, N.Y.
Halsey, S.D. (1978). Late Quaternary geologic history and morphologic development of the barrier island system along
the Delmarva Peninsula of the mid-Atlantic Bight. Ph.D.
dissertation, Univ. of Delaware, 592 p.
Halsey, S.D., Farrell, S.C., Hammond, J.J., and Kraft, J.C.
(1977). Preliminary investigations of former coastal features preserved along the mid-Wisconsinan(?) shoreline of
New Jersey and Delmarva. Geol. Soc. America, Abs. with Programs (Northeastern Sec.), v. 9, no. 3, p. 46.
Halsey, S.D., Farrell, S.C., and Johnson, S.W. (1979). Further
investigations of the geomorphic history of the mid-Wiscon-
sinan(?) coastal system of New Jersey. Geo. Soc. America,
Abs. with Programs (Northeastern Sec.),v. 11, no. 1, p.15-16.

Hayes, M.O., (this volume). A comparison of microtidal and meso-
    tidal barrier island systems.
Hayes, M.O. and Ruby, C. (1975). Barrier island development on
    a tectonically active delta, Copper River delta, Alaska.
    Geol. Soc. America, Abs. with Programs (Ann. Mtg.), v. 7,
    no. 7, p. 1105-1106.
Hoyt, J.H. (1967). Barrier island formation. *Geol. Soc. America
    Bull. 78*, p. 1125-1136.
Hoyt, J.H. and Hails, J.R.(1974). Pleistocene stratigraphy of
    southeastern Georgia. *In* "Post-Miocene stratigraphy central
    and southern, Atlantic Coastal Plain (R.Q. Oaks and J.R.
    DuBar, eds.), p. 191-205. Utah State Univ. Press, Logan.
Inman, D.L. and Nordstrom, C.F. (1971). On the tectonic and
    morphological classification of coasts. *Jour. Geology 79*,
    p. 1-21.
Kaczorowski, R.T. (1975). Offset tidal inlets, Long Island,
    New York. Geol. Soc. America, Abs. with Programs (North-
    eastern/Southeastern Sec.), v. 8, no. 2, p. 207.
Kraft, J.C. (1971a). Sedimentary facies patterns and geologic
    history of a Holocene marine transgression. *Geol. Soc.
    America Bull. 82*, p. 2131-2158.
Kraft, J.C. (1971b). A guide to the geology of Delaware's
    coastal environments. Guidebook, Geol. Soc. America.
    College of Marine Studies Publ., Univ. of Delaware, 220 p.
Kraft, J.C., Biggs, R.B., and Halsey, S.D. (1973). Morphology
    and vertical sedimentary sequence models in Holocene trans-
    gressive barrier systems. *In* "Coastal Geomorphology" (D.R.
    Coates, ed.), p. 321-354. Publ. in Geomorphology, Bing-
    hamton, N.Y.
Leatherman, S.P. (1976). Quantification of overwash processes.
    Ph.D. dissertation, Univ. of Virginia, 245 p.
Moslow, T.F. and Heron, S.D. Jr. (1977). Evidence of relict
    inlets in the Holocene stratigraphy of Core Banks from
    Cape Lookout to Drum Inlet. Geol. Soc. America, Abs. with
    Programs (Southeastern Sec.), v. 9, no. 2, p. 170.
Moslow, T.F. and Heron, S.D. Jr. (1978). Recent evolution of
    Cape Lookout, North Carolina. Geol. Soc. America, Abs. with
    Programs (Southeastern Sec.), v. 10, no. 4, p. 193.
Moslow, T.F. and Heron, S.D. Jr. (this volume). Quaternary
    evolution of Core Banks, North Carolina: Cape Lookout to
    New Drum Inlet.
Otvos, E.G. Jr. (1970). Development and migration of barrier
    islands, Northern Gulf of Mexico. *Geol. Soc. America Bull.
    81*, p. 241-246.
Otvos, E.G. Jr. (1977). Post-Pleistocene history of the United
    States inner continental shelf: significance to origin of
    barrier islands: Discussion. *Geol. Soc. America Bull. 88*,
    p. 734-736.

Owens, J.P. and Denny, C.S. (1974). Provisional stratigraphic sequences in the Maryland-Delaware parts of the lower Delmarva Peninsula. Geol. Soc. America, Abs. with Programs (Northeastern Sec.) v. 5, no. 1, p. 61-62.

Pierce, J.W. and Colquhoun, D.J. (1970). Holocene evolution of a portion of the North Carolina coast. *Geol. Soc. America Bull. 81*, p. 3697-3714.

Rampino, M.R. (1973). Geology of the south shore of Long Island and the history of the late Pleistocene glaciation. (Abs.) Proc. 6th Ann. Long Island Sound Conf., p. 31.

Rampino, M.R. and Sanders, J.E. (1976). New stratigraphic evidence for major mid-Wisconsinan climatic oscillation and sealevel rise: "Interstadial" or Interglacial"?: Geol. Soc. America, Abs. with Programs (Ann. Mtg.), v. 8, no. 6, p. 1059-1060.

Rampino, M.R. and Sanders, J.E. (1977). Modern barrier islands built atop Late Pleistocene barriers: historical implications. Geol. Soc. America, Abs. with Programs (Northeastern Sec.), v. 9, no. 3, p. 310-311.

Riggs, S.R. and O'Connor, M.P. (1975). Evolutionary succession of drowned Coastal Plain estuaries. Geol. Soc. America, Abs. with Programs (Ann. Mtg.), v. 7, no. 7, p. 1247-1248.

Schwartz, M.L. (1971). The multiple causality of barrier islands. *Jour. Geology 79*, p. 91-94.

Schwartz, M.L. (ed.)(1973). Barrier Islands: Benchmark Papers in Geology, v. 9. Dowden, Hutchinson and Ross, Stroudsburg, Pa., 451 p.

Shepard, F.P. and Wanless, H.R. (1971). Our changing coastlines. McGraw Hill, New York, 579 p.

Sheridan, R.E., Dill, C.E. Jr., and Kraft, J.C. (1974). Holocene sedimentary environments of the Atlantic inner shelf off Delaware. *Geol. Soc. America Bull. 85*, p. 1319-1328.

Susman, K.R. and Heron, S.D. (1976). Post-Miocene subsurface stratigraphy of Shackleford Banks, a barrier island near Cape Lookout, N.C. Geo. Soc. America, Abs. with Programs (Northeastern/Southeastern Sec.), v. 8, no. 2, p. 280-281.

Susman, K.R. and Heron, S.D. (1979). Evolution of a barrier island, Shackleford Banks, Carteret County, North Carolina. *Geol. Soc. America Bull.*, Part I, v. 90, p. 205-215.

Swift, D.J.P. (1969). Inner shelf sedimentation: processes and products. *In* "The NEW concepts in continental margin sedimentation" (D.J. Stanley, ed.), p. DS-4: 1-46. American Geol. Institute, Washington, D.C.

Swift, D.J.P. (1975). Barrier island genesis: evidence from the central Atlantic shelf, eastern U.S.A. *Sed. Geology 14*, p. 1-43.

Truitt, R.V. (1968). High winds...high tides: a chronicle of
    Maryland's coastal hurricanes. Ed. Ser. No. 77, Natural
    Resources Institute, Univ. of Maryland, 35 p.

Truitt, R.V. (1971). Assateague...the "place across": a saga
    of Assateague Island. Ed. Ser. No. 90, Natural Resources
    Institute, Univ. of Maryland, 48 p.

Webster (1963). Seventh New Collegiate Dictionary. Merriam Co.,
    Springfield, Mass., 1221 p.

Wroten, W.H. Jr. (1970). Assateague. Peninsula Press, Salisbury,
    Maryland, 46 p.

Zeigler, J.M. (1959). Origin of the Sea Islands of the south-
    eastern United States. *Geog. Rev. 49*, p. 222-237.

QUATERNARY EVOLUTION OF CORE BANKS, NORTH CAROLINA:
CAPE LOOKOUT TO NEW DRUM INLET

Thomas F. Moslow

Department of Geology
University of South Carolina
Columbia, South Carolina

S. Duncan Heron, Jr.

Department of Geology
Duke University
Durham, North Carolina

Forty-six auger and wash-bore holes were drilled along a 36 km segment of Core Banks, North Carolina, to examine the Quaternary geologic history of a typical mid-Atlantic barrier island. Particular attention has been focused on the Holocene evolution of Core Banks including its origin, landward migration and overall response to a transgressing sea.
 Six lithologic units were identified. These are:

 (1) the Yorktown Formation (early Pliocene), a green-gray sandy clay
 (2) a 2.4 m to 3.1 m thick case-hardened limestone cap composed of diagenetically altered Yorktown sediments
 (3) the Core Creek sand (late Sangamon), a transgressive nearshore-marine facies
 (4) the Atlantic sand (mid Wisconsin), a transgressive barrier island complex
 (5) the Diamond City Clay (late Wisconsin), a regressive lagoonal sequence
 (6) the Core Banks sand (Holocene), consisting of transgressive barrier, backbarrier, and inlet-fill sands.

Nine depositional environments were identified within the Holocene sediments. In five isolated sections the Holocene stratigraphy is completely reworked by tidal inlets that opened, migrated and closed sometime during the past 4,000 years. Arcuate, relict flood tidal deltas on the soundside of Core Banks are found in association with each of these inlets.

The southernmost four to five km of Core Banks, including Cape Lookout, has been formed by spit accretion during the past 4,000 years. Here a regressive sequence of overwash sands over-lying shoreface deposits occurs within the Holocene.

Core Banks probably originated as part of an elongating spit, or by mainland beach detachment on the nearshore shelf about 15,000 years ago. From this time and position, Core Banks has steadily migrated landward in response to a rising sea. Rates of landward migration were calculated using Holocene C-14 dates and projected locations of paleoshorelines. Paleoshorelines were located using the assumption that the entire shoreface has moved with the rising sea level according to Brunn's Rule and the Swift concept of shelf evolution. Core Banks has migrated about 6.7 km landward in the past 7,000 years. From 7,000-4,000 BP the island migrated at rates ranging from 98 m to 45 m per century. This was a period of minor inlet deposition and a dominance of oceanic overwash. The rate of landward migration decreased markedly in the interval from 4,000 BP to 755 BP, reflecting a slackening in sea level rise. During this portion of the Holocene record, inlet formation and migration are dominant.

INTRODUCTION

Scope

This study was conducted to examine the pre-Holocene and Holocene deposits beneath Core Banks along two main lines of investigation. The first was to correlate the Plio-Pleistocene stratigraphy beneath Core Banks to the transgressions and regressions of the sea which have dominated the post-Miocene depositional history of the North Carolina Coastal Plain. Secondly, the Holocene stratigraphy of Core Banks was examined for the purpose of determining the dynamic processes that originally formed and continue to mold the island. Particular attention has been focused on the Holocene evolution of Core Banks including its origin, landward migration, and overall response to a transgressing sea.

Geologic Setting

Core Banks, including Portsmouth Island, is approximately 64 km long, extending northeast from Cape Lookout to Ocracoke Inlet, and forms the southern leg of the Outer Banks of North Carolina (Fig. 1). Core Banks is bordered to the east by the Atlantic Ocean and separated from the mainland by the shallow, narrow Core Sound. This lagoon is approximately 4.8 km wide and has an average depth of only 1.6 m. The study area consists of the southern half of the Core-Portsmouth Banks chain. Referred to simply as "Core Banks," the island is approximately 36.2 km long, extending from Cape Lookout to New Drum Inlet.

Core Banks has an ocean tidal range of approximately 0.9 m with a morphology typical of a high wave energy, microtidal barrier island with its elongate shape, infrequent tidal inlets, large flood-tidal deltas and abundance of washover fans and terraces (Hayes 1979). The island has an average elevation of 1.5 - 3.0 m and an average width of only 0.8 km (Fig. 2).

Core Banks is frequently subjected to large amounts of oceanic overwash; sand is often transported completely across the island and into the sound. Island breaching from storm surges has resulted in the opening of at least nine inlets between Ocracoke and Beaufort since 1585 (Fisher 1962). Deposition from these relict inlets has left large, arcuate flood-tidal deltas along the back side of Core Banks. This can be readily seen in Figure 3. According to Godfrey (1970), extensive salt marsh develops on these former flood-tidal deltas and are in turn buried by overwash fans. This overwash process, in conjunction with shoreline erosion from a rising sea level, results in the landward migration of Core Banks.

Methods

A total of three wash-bore and 43 auger holes were drilled on Core Banks from Cape Point to New Drum Inlet (Fig. 4). Wash-bore samples were taken with a Lynac split-spoon (sample size 3.8 x 61 cm) driven into fresh sediment using a 63.5 kg safety hammer. The remaining 43 holes were drilled by power augering. All drilling was done with a truck-mounted Mobil Drill rig. Fifteen holes penetrated to an early Pliocene basement. All 46 were drilled through the Holocene.

Location of drill sites along Core Banks was limited to the upper berm, dune ridge and overwash passes, and backbarrier flat portions of the island. The excessive weight and lack of four-wheel drive on the drill rig prevented work on the beach or in the intertidal marsh environments. Therefore, the only cross-section made across the width of the barrier was at Cape

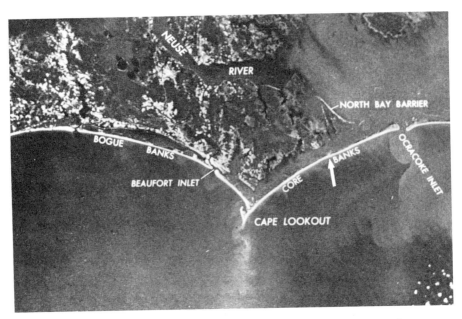

*Fig. 1.   Apollo 9 photograph of the Core Banks region,
North Carolina. This investigation was conducted on the
southern portion of Core Banks from Cape Lookout to Drum Inlet
(arrow). The northern half of Core Banks is also referred to
as Portsmouth Banks and extends to Ocracoke Inlet. (Photograph
by NASA, March 1969, from Mixon and Pilkey, 1976.)*

Lookout (Fig. 4). However, in order to display the probable relationship of depositional facies perpendicular to the shoreline, a hypothetical cross-section was constructed (Figs. 6 and
9).

PRE-HOLOCENE STRATIGRAPHY AND DEPOSITIONAL HISTORY

Six lithologic units are recognized within the post-Miocene
stratigraphy of Core Banks extending over the past seven million
years of geologic history along the southeastern Coastal Plain
and nearshore shelf of North Carolina. These units were defined
by lithology, texture, and faunal assemblages (Table 1). A post-
Miocene cross-section of Core Banks displays those units and
associated C-14 dates (Fig. 5).

TABLE I. *Sedimentary Characteristics of Pre-Holocene Depositional Units*

Unit/Age	Color	Lithology	Molluscs	Depositional Environment
Diamond City Clay (late Wisconsin)	dark gray	silty-clay	*Mulinia* sp. *Crassostrea* sp.	Estuary/ Lagoon
Atlantic sand (mid-Wisconsin)	tan-gray	well-sorted, clean VF-C sand	*Donax* sp. *Mulinia* sp. *Crassostrea* sp.	Barrier complex
Core Creek sand (Sangamon)	gray-green	silty-sand & clay	*Mercenaria* sp. *Mulinia* sp.	Nearshore marine
Yorktown Fm. (early Pliocene)	pale green	sandy-clay	*Ecphora* sp. *Chlamys* sp.	Nearshore marine

Fig. 2. *Oblique aerial photograph of a portion of Core Banks showing a morphology typical of microtidal barrier islands. View is to the west towards Core Sound.*

Fig. 3. (a) Aerial view of the active flood-tidal delta at
New Drum Inlet. This sand body is very similar in size and
shape to the relict flood-tidal deltas along the sound side of
Core Banks. (b) Aerial photograph of the relict flood-tidal
delta associated with the largest and deepest inlet-fill body
on Core Banks (cross-section A-A'). View is to the west looking across the barrier into Core Sound.

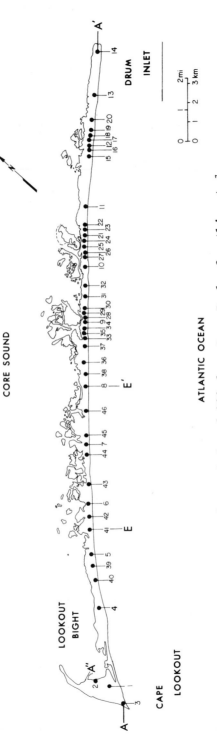

Fig. 4. Location map for the 46 holes drilled on Core Banks for this study.
Cross-section end points are indicated by capital letters.

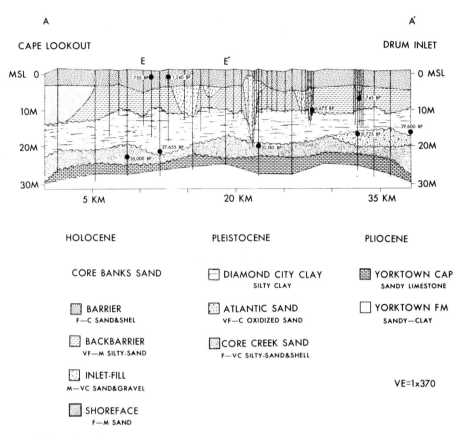

Fig. 5.   CROSS-SECTION A-A', QUATERNARY OF CORE BANKS.
All but the two Pliocene stratigraphic units are separated by
unconformities.

Pliocene

The uppermost Tertiary strata is represented by the unconsolidated sediments of the Yorktown Formation. Beneath Core
Banks this unit is typically a green-gray, well sorted, very
fine to fine-grained clayey sand. Mollusc assemblages imply a
nearshore marine or shallow shelf environment of deposition during the late Yorktown Transgression of the sea (Table I).
During early Pliocene time, a period of regional emergence
of the North Carolina Coastal Plain ensued. Subaerial exposure
resulted in erosion and diagenetic alteration of the Yorktown
beds. This event is represented beneath Core Banks by a highly

altered packstone forming the upper 2.4 m to 3.1 m of Yorktown sediments (Fig. 5). The unconformity between this limestone cap unit and the overlying beds marks a major stratigraphic break representing all of early Pliocene to late Pleistocene time (over 6 million years). Subaerial exposure and nearshore marine erosional processes have apparently removed any record of deposition during this time. Early and medial Pleistocene transgressions, recorded elsewhere on the North Carolina Coastal Plain, are missing beneath Core Banks.

Pleistocene

Deposition in the Core Banks area during late Pleistocene time is recorded by three distinct stratigraphic units. The oldest of these is the Core Creek sand, a lithologically varied unit which is dominantly a silty and clayey, highly fossiliferous, fine- to coarse-grained, quartz sand. In late Pleistocene time the Core Creek sand was deposited regionally by a late Sangamon transgression of the sea 80,000 - 120,000 BP (Mixon and Pilkey, 1976; Sussman and Heron, 1979). This sea advanced about 32 km inland from its present position giving rise to it highest shoreline, the Arapahoe Ridge (Daniels et al., 1977). The nearshore marine and tidal delta facies of the Core Creek sand that occur beneath Core Banks were deposited seaward of this barrier/ridge as a terrace formation. The upper beds of the Core Creek found elsewhere are missing beneath Core Banks, evidence of their removal by erosion processes during the early Wisconsin regression of the sea.

A Pleistocene barrier complex, the Atlantic sand, was found in the four northernmost deeper holes drilled on Core Banks from -14.6 m to -19.8 m MSL (Fig. 5). This unit consists of very fine- to coarse-grained, well sorted, clean, quartz sands, and was deposited during the latter part of the mid-Wisconsin transgression before 35,000 BP (Mixon and Pilkey, 1976). The Atlantic sand was probably part of a transgressing barrier chain that bordered the open ocean. Mollusc assemblages found within this unit represent back-barrier and barrier shoreface environments (Table I). The Atlantic sand also occurs on the coastal plain adjacent to Core Sound (Fig. 6). These mainland exposures of the Atlantic sand and those found beneath Core Banks are in a northwest alignment that fits the postulated configuration of the Pleistocene barrier island chain of Pierce and Colquhoun (1970). They suggest that this portion of the Pleistocene barrier chain represents an ancestral Cape Lookout.

The Diamond City Clay is the youngest Pleistocene unit. These lagoonal silty-clays were deposited during the late Wisconsin regression (Susman and Heron, 1979). A lagoon much wider and deeper than present-day Core Sound probably existed in the

Table II. Radiocarbon Dates

Ref. # Univ. of Miami	Hole #	Depth (m/MSL)	Material	Age (years BP)	Remarks
			CORE BANKS SAND		
UM-1097	43	-0.3 m	Spartina peat in a sandy mud matrix	1,240 ± 125	Backbarrier intertidal salt marsh bed
UM-917	42	-0.6 m	Peat and root fragments in a tidal marsh mud	755 ± 100	0.26 ft. (.08 m)/century sea-level rise
UM-907	17	-6.4 m	Peat and root fragments in a tidal marsh mud	5,745 ± 215	0.36 ft (0.11 m)/century sea-level rise
UM-1098	19	-7.7 m	Shells, possibly in growth position, in a backbarrier sandy mud	4,265 ± 115	Biocoenosis (in place life assemblage)
UM-908	23	-9.5 m	Peat and root fragments at base of Holocene section	6,675 ± 160	0.45 ft (0.13 m)/century sea-level rise
			DIAMOND CITY CLAY		
UM-853	6	-21.5 m	Wood	$27{,}655 \, {}^{+945}_{-1070}$	Salt marsh peat bed
UM-906	14	-17.1 m	Peat with wood	29,600 ± 140	Salt marsh peat bed
			ATLANTIC SAND		
UM-905	12	-15.7 m	Peat with wood	31,725	Salt marsh
			CORE CREEK SAND		
UM-859	9	-18.9 m	Wood (some peat)	32,185	Salt marsh peat bed
UM-857	5	-21.9 m	Peat with some wood	$35{,}000 \, {}^{+1125}_{-1325}$	Salt marsh peat bed
UM-858	5	-21.9 m	Peat with some wood	31,295	(both dates of same material)

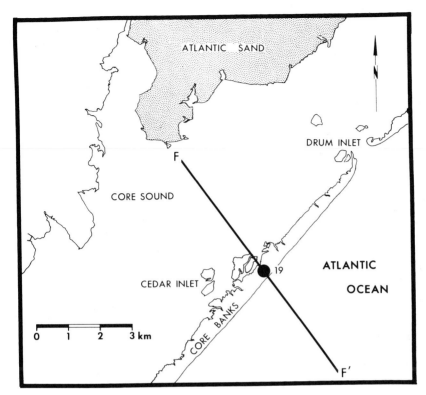

Fig. 6. Location map for cross-section F-F' across the width of the barrier near the northern end of Core Banks. The mainland occurrence of the Atlantic Sand is shown by the stippled pattern.

study area at this time. Any barrier complex associated with this lagoon was either eroded or prograded out onto the near-shore shelf. C-14 dates indicate that the Diamond City was deposited beneath Core Banks from about 24,000 to 29,000 BP (Table II). This regressive lagoonal sequence occurs at -9.0 m to -10.0 m MSL, except where the Pleistocene beds are truncated by the scouring effects of Holocene migrating inlets (Fig. 5).

HOLOCENE STRATIGRAPHY AND EVOLUTION OF CORE BANKS

Sedimentary Units

Unconformably overlying the Pleistocene lagoonal deposits along the entire length of Core Banks is a diverse sequence of Holocene barrier deposits. The average depth of occurrence for the Holocene/Pleistocene contact is generally about -9 m MSL. The sequence of Holocene sediments averages from 10-12 m in thickness.

Holocene sediments beneath Core Banks have been divided into three depositional complexes; barrier, backbarrier, and migrating inlet. These typically fossilliferous, fine- to coarse-grained, tan to light gray sands and silts are represented by nine different depositional environments. A description of the characteristic sediments associated with each depositional environment is given in Table III.

Stratigraphic Relationships

The Holocene sediments beneath Core Banks reveal a complex depositional history dominated by barrier retreat, spit extension and inlet migration. A transgressive sequence of sediments similar to that underlying the Delmarva Peninsula (Kraft 1971) dominates the Holocene stratigraphy. The common occurrence of barrier overwash sands overlying backbarrier silty-sands and salt marsh peats indicate that landward migration has been an active process in the island's evolution. However, in five isolated sections, the Holocene section has been completely reworked by the action of migrating tidal inlets (Fig. 5).

The five relict inlets found beneath Core Banks are typically represented by three depositional facies (Table III). These are: (1) inlet floor, (2) main channel, and (3) inlet margin (spit platform). The stratigraphic relationship and relative thickness of the inlet floor and main channel facies is shown in Fig. 7. The inlet margin facies is not preserved in all instances. The inlet floor and main channel sediments form the migrating inlet proper and represent the bulk volume of inlet-fill. Approximately 15% of the Holocene sediments beneath Core Banks are inlet-related deposits (Moslow and Heron, 1978). The inlet-fill deposits displace Holocene backbarrier silty-sands and are overlain by medium- to coarse-grained washover sands (Fig. 7).

The stratigraphic relationship of all three barrier-type depositional units are best shown in the three wash bore holes drilled at Cape Lookout (Fig. 8). The Holocene section is a regressive sequence of coarse-grained, interlaminated washover sands resting on fine-grained, burrowed, shoreface deposits. A major facies change occurs within the Holocene section somewhere between drill holes 1 and 4 (Fig. 4). Here the open-marine shoreface deposits found beneath Cape Lookout displace the backbarrier silty-sands found in the same stratigraphic position beneath the remainder of Core Banks (Fig.5).Apparently, the southernmost four to five km of present-day Core Banks, including Cape Lookout, has formed by spit progradation. This

TABLE III. *Holocene Depositional Environments*

Complex	Depositional Environments	Sediment Characteristics
	Overwash	Medium to coarse sand and shell
Barrier	Shoreface	Fine sand, well sorted
	Recurved spit	Coarse sand, poorly sorted
Backbarrier	Flood-tidal delta	Fine to coarse sand, silt
	Backbarrier lagoon	Silty-sand, well sorted
	Salt marsh	Fine sand, silt and clay; plant remains
Migrating Inlet	Inlet margin	Fine sand, well sorted
	Main channel	Coarse sand and abraded shells
	Inlet floor	Shell and pebble lag

southward extension is a result of the combined effects of washover deposition and a southerly longshore transport (Fisher 1968; Field and Duane, 1976).

The fine- to very coarse-grained, poorly sorted sand and shell in drill hole two (Fig. 8) was deposited by the progradation of a recurved spit on the landward side of Cape Lookout (Fig. 1).

In order to depict the relationship of barrier and backbarrier depositional facies across the width of Core Banks, cross-section F-F' was constructed (Figs. 6,9). Although based on the sediments retrieved in one drill hole, this hypothetical cross-section displays the probable geometry and profile of the Holocene barrier perpendicular to the shoreline. The barrier sands that cap the Holocene section in Fig. 9 interfinger with shoreface and flood-tidal delta sediments seaward and landward of the barrier.

Rates of Landward Migration

Introduction. The ongoing Holocene transgression of the sea has been the dominant factor in determining Core Banks' mode of formation and continued evolution. This is apparently true for many other barrier island systems along the mid-Atlantic and Gulf coasts (Hoyt 1967; Otvos 1970; Kraft 1971; Swift 1975; Field and Duane, 1976; Susman and Heron, 1979).

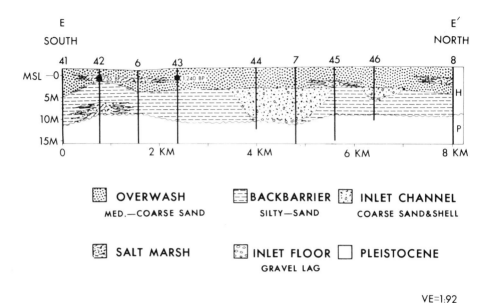

VE=1.92

Fig. 7. SECTION E-E' ZACK CREEK. Cross-section E-E' show-ing a portion of the Holocene stratigraphy. Cross-section loca-tion is shown in Figure 4 (from Moslow and Heron, 1978).

In response to a rising sea level, Core Banks has moved across the continental shelf to its present position. The record of this migration is documented within the Holocene sediments be-neath Core Banks. Stratigraphic sequences and primarily radio-carbon dates have been used to determine the total extent and rate of this landward translation.

The Theory. No generally accepted sea-level curve exists specifically for the North Carolina coast. However, numerous investigations have been conducted on the Holocene transgression of the sea along the mid-Atlantic east coast of the United States (Curray 1965; Milliman and Emery, 1968; Bloom 1970; Kraft 1971; Emery and Uchupi 1972; Kraft 1976; Belknap and Kraft, 1977). Most of these studies indicate a marked decrease in the rate of sea-level rise at about 7,000 years BP. The three curves shown in Fig. 10 reflect late Holocene sea level along the United States east coast. All three depict a steady transgression of the sea at this time, while the upper two curves show a subtle deceleration of sea-level rise at about 4,000 BP.

Bruun (1962) was perhaps the first to examine the quanti-tative effect a slowly transgressing sea has on a barrier

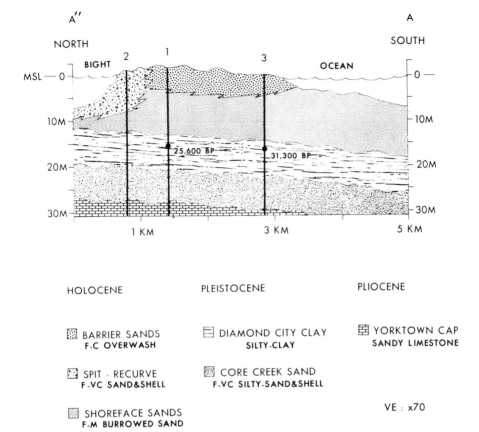

NORTH   SOUTH

HOLOCENE   PLEISTOCENE   PLIOCENE

⊞ BARRIER SANDS   ⊟ DIAMOND CITY CLAY   ⊞ YORKTOWN CAP
   F-C OVERWASH   SILTY-CLAY   SANDY LIMESTONE

⊡ SPIT - RECURVE   ⊡ CORE CREEK SAND
   F-VC SAND&SHELL   F-VC SILTY-SAND&SHELL

⊡ SHOREFACE SANDS   VE : x70
   F-M BURROWED SAND

*Fig. 8.   QUATERNARY OF CAPE LOOKOUT.   Cross-section A″-A
of Cape Lookout. Holocene shoreface sands completely displace
backbarrier deposits. Note the occurrence of interfingering
recurved spit deposits in hole 2. The carbon-14 dates were
provided by the Shell Development Company.*

shoreline (Fig. 11). The shoreline erodes at a rate which is in
equilibrium with vertical accretion and deposition of the shoreface. This is reflected in the change in configuration of the
shoreface from the initial to the resulting profile in Fig. 11.
As a result, the same depth of water is maintained over the
shoreface. The overall response of a Core Banks type of barrier
to rising sea level is shoreline erosion and consequent landward migration through time (Fig. 12).

The diagrammatic cross-section in Fig. 13 depicts successive
positions of the shoreface during the late Holocene transgression. The upper half of the diagram shows how the combination

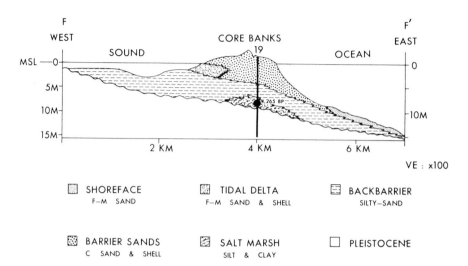

*Fig. 9. HOLOCENE BARRIER PROFILE. Holocene cross-section
F-F' showing facies relationships across the width of the
barrier.*

of periods of stillstand and transgression creates terraces
with seaward facing scarps on the shoreface. The resulting
stratigraphy displayed in the lower diagram depicts a transgressive type of barrier island identical in nature to Core
Banks.

The Model. In determining specific rates of migration for
Core Banks it is first assumed that the depth of occurrence of
the Holocene C-14 material is roughly equal to the relative
stand of sea level at the time equal in age (years BP) to the
date of the material. Four of the five Holocene C-14 dates
(Table II) are of salt marsh Spartina peats, which are generally
considered to be fairly accurate sea-level indicators (Kraft
1971; Belknap and Kraft, 1977).

     Therefore, if the shoreface has maintained its equilibrium
profile through time and the Holocene C-14 dates are accurate
sea-level indicators, then paleo-shorelines for Core Banks
should be situated approximately at the distance offshore at
the contour equal to the depth of occurrence of the dated material. For instance, material dated at -9.5 m MSL (Table II)
indicates that the paleo-shoreline at the time of the C-14 date
(6,675 BP) is approximately 6.5 km offshore where the water
depth is -9.5 m (Fig. 14). The diagram in Figure 14 was constructed to attempt to determine the distance offshore of

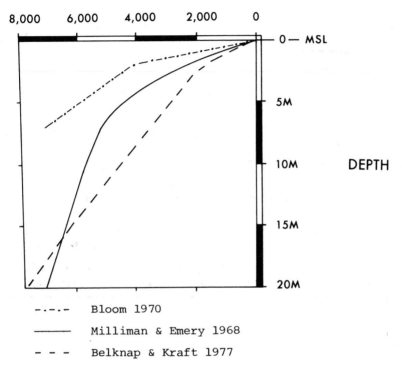

AGE : YRS. BP

- · - · -   Bloom 1970

————   Milliman & Emery 1968

- - -   Belknap & Kraft 1977

Fig. 10.   LATE HOLOCENE SEA LEVEL.   Curves reflecting late Holocene sea-level history. The curves of Bloom (1970), and Milliman and Emery (1968) represent eustatic sea level while that of Belknap and Kraft (1977) represent relative sea-level rise.

paleo-shorelines for Core Banks. These distances should theoretically represent the extent of landward migration of Core Banks through Holocene time. The distance offshore divided by time of occurrence of the paleo-shoreline from present day Core Banks yields a rate of migration. The computed rates are plotted against late Holocene time in Figure 15. The curve indicates that the rate of landward migration decreases sharply with time during the Holocene as would be exptected by a slower rise in sea level to the present. The lack of Holocene C-14 dates obviously makes the curve rather speculative, but it is felt that this model can be used to determine rates of barrier migration in relation to the island's depositional history.

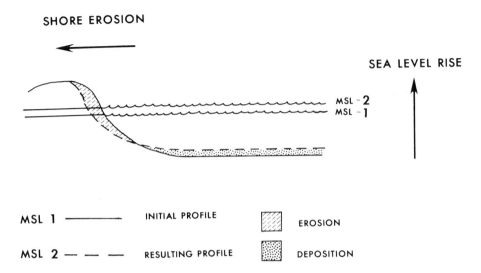

Fig. 11.   Diagram displaying the equilibrium profile between
shore erosion and nearshore/shoreface deposition (Brunn 1962).

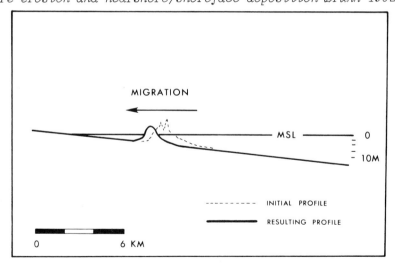

Fig. 12.   BARRIER RESPONSE TO SHORELINE EROSION.
Diagram showing the response of a barrier island to shoreline
erosion from a rise in sea level (from Hoyt 1967).

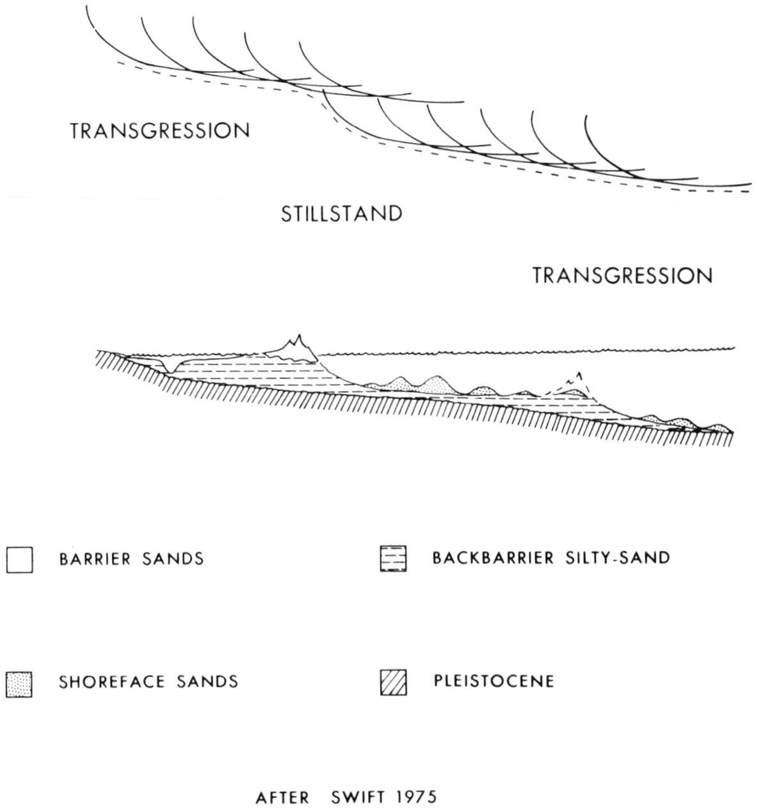

TRANSGRESSION

STILLSTAND

TRANSGRESSION

☐ BARRIER SANDS   ☰ BACKBARRIER SILTY-SAND

▦ SHOREFACE SANDS   ▨ PLEISTOCENE

AFTER SWIFT 1975

Fig. 13. BARRIER SHORELINE RETREAT. Diagram depicting successive positions of the shoreface during transgression and stillstand of the late Holocene. The resulting stratigraphy is shown in the lower half of the diagram (from Swift 1975).

Effects on Depositional Patterns. Results indicate that Core Banks has migrated landward as much as 6.7 km over the past 6,675 years (Fig. 16). From about 7,000 BP to 4,000 BP the island was migrating at rates averaging between 45 m - 98 m per century. This rapid landward retreat probably resulted in near constant oceanic washover with little or no dune development along the island. Washover fans predominated and the fringing marsh in the lagoon was very well developed as depicted in Figure 16. There were probably many small breaches along Core Banks. However, it is doubtful that any hydraulically active

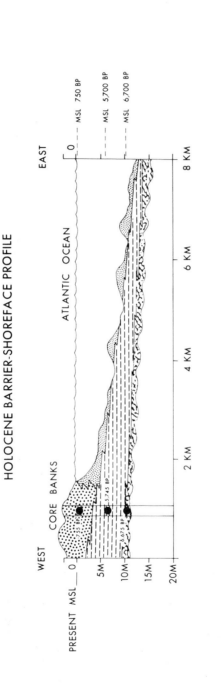

Fig. 14. Holocene profile of Core Banks and the adjacent shoreface. This composite cross-section contains three C-14 dates from various drill holes within the Holocene stratigraphy. The former stands of sea level shown along the right hand column are based on the model presented here.

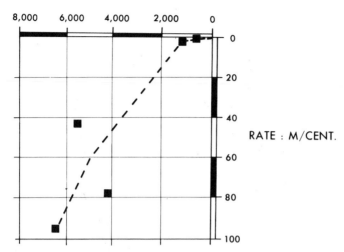

AGE : YRS. BP

*Fig. 15.   LANDWARD MIGRATION CURVE: CORE BANKS.   Curve dis-
playing the landward migration rates of Core Banks for about
the past 7,000 years. Lack of more C-14 dates makes the curve
obviously conjectural. However, it can be used to roughly
approximate rates of landward migration for this barrier island
shoreline.*

tidal inlets existed at this time. Inlets were shallow and
ephemeral, rapidly opening and closing, leaving no discrete
inlet channels owing to the vast amount of overwash.

The subtle decrease in the rate of sea-level rise about
4,000 years ago (Fig. 10) significantly affected the behavior
and geomorphology of Core Banks (Fig. 16). This is the second
and apparently the last sharp break in sea-level rise recorded in the Holocene beneath the island. The sharp decline in
the rate of landward migration of Core Banks at about this
time (Fig. 15) reflects this drop in sea-level rise. Inlet formation and migration dominate this portion of the Holocene record.

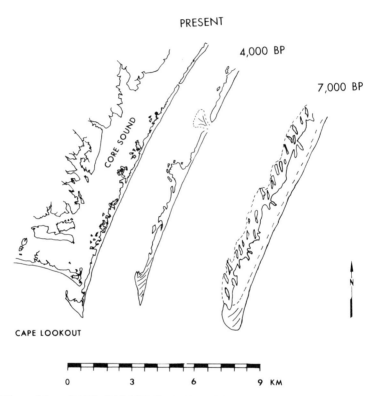

Fig. 16. LATE HOLOCENE EVOLUTION OF CORE BANKS. Stages in the late Holocene evolution of Core Banks. The diagram shows the relative distance offshore, and paleogeomorphology of the barrier at 7,000 and 4,000 BP based on computed rates of landward migration.

Other probable effects on the island's behavior at this time were decreased amounts of oceanic overwash and the formation of a few short-lived, yet hydraulically active, tidal inlets. These inlets opened, migrated laterally and closed sometime during the past 4,000 years. Within the Holocene stratigraphy beneath Core Banks there is evidence (Moslow and Heron, 1978) for five such relict inlets (Fig. 5).

The decreased rate of landward migration, the landward extension of washover fans and the development of large flood-tidal deltas were active processes along Core Banks during the past 4,000 years (Fig. 16). The decreased rate of sea-level rise and the increase in sedimentation from washover and tidal exchange resulted in a shallowing of Core Sound. This infilling of Core Sound has resulted in a decrease in the number of inlets along Core Banks. Core Sound today has an average depth of only 1.6 m. Along all of Core and Portsmouth Banks only the New Drum Inlet is active today (Fig. 1). Even this inlet was artificially opened and requires extensive maintenance through dredging to keep it from filling in.

As Core Banks became relatively stable in respect to sea level rise, large volumes of sand were being transported alongshore to Lookout Shoals. The slower retreat of the island probably induced the southerly extension of Core Banks (Fig. 16) along with a buildup of Cape Lookout Shoals at this time (Fisher 1968).

CONCLUSIONS

Analysis of the sediments from 46 wash-bore and auger holes drilled on Core Banks from Cape Lookout to Drum Inlet has led to the following conclusions:

(1) Uppermost Tertiary strata beneath Core Banks is represented by beds of the Yorktown Formation (early Pliocene). These sediments were deposited by the late Yorktown transgression of the sea.

(2) During early Pliocene time a eustatic drop in sea level resulted in emergence of the North Carolina Coastal Plain. This event is represented beneath Core Banks by a highly altered packstone forming the upper 1.5 to 3.0 m of Yorktown sediments.

(3) A major stratigraphic break occurs representing all of early Pliocene to late Pleistocene time (over 6 million years). Subaerial exposure and nearshore marine erosional processes have apparently removed any record of deposition during this time.

(4) In the late Pleistocene the nearshore marine and tidal-delta sediments of the Core Creek sand were deposited in a late Sangamon sea. This sea advanced about 32 km inland from Core Banks, giving rise to its highest shoreline, the Arapahoe Ridge.

(5) The Atlantic sand beneath Core Banks was deposited as part of a Pleistocene barrier island complex that extended out onto the continental shelf during the later part of the mid-Wisconsin transgression, and may represent an ancestral Cape Lookout.

(6) As sea level fell during the late Wisconsin regression, the lagoonal silty-clay beds of the Diamond City were deposited. A lagoon much wider and deeper than Core Sound existed in the study area at this time.

(7) The Holocene sequence beneath Core Banks records the last 7,000 years of geologic history. It is represented by at least nine different depositional environments.

(8) Core Banks probably originated as part of an elongating spit complex or by mainland beach detachment far out on the nearshore shelf about 15,000 years ago.

(9) Since that time, it has migrated landward to its present position in response to a rising sea. About 7,000 years ago, Core Banks was about 6.7 km offshore from its present position. From 7,000 to 4,000 BP the island migrated at rates between 45 m to 98 m per century. Since this time, the island's landward migration decreased drastically.

(10) At least five tidal inlets have opened, migrated, and closed along the Core Banks study area. This occurred definitely within the last 4,000 years and possibly during the past several hundred years.

(11) Over the past 4,000 years, Core Banks has migrated only about 1.6 km to 3.2 km landward (two or three times its own width). This relative stability allowed for the development and extension of Cape Lookout. The southernmost 4.0 - 5.0 km length of Core Banks was formed by spit elongation through longshore transport.

(12) The following processes have dominated Core Banks morphology over the past 4,000 years:
  (a) decreased rates in landward migration
  (b) landward extension of washover fans
  (c) development of large flood-tidal deltas from inlet deposition.

(13) The decreased rate of sea-level rise and the increase in sedimentation from washover and tidal exchange has resulted in a shallowing of Core Sound. This infilling of Core Sound has resulted in a decrease in the number of inlets along Core Banks today.

ACKNOWLEDGMENTS

Financial assistance for this project was funded by a grant from the National Park Service to Duke University. O.H. Pilkey, R.D. Perkins, and Ken Susman are acknowledged for their valuable discussions concerning the stratigraphy and evolution of Core Banks. Blake Blackwelder and William Miller identified numerous mollusc assemblages and interpreted their environments of

deposition. Coral fragments were identified by Robin Lighty.
Jude Wilber and Chris Osborn provided valuable assistance in
the field.

REFERENCES CITED

Belknap, D.F. and Kraft, J.C. (1977). Holocene relative sea-
   level changes and coastal stratigraphic units on the north-
   west flank of the Baltimore Canyon Trough geosyncline. *Jour.
   Sed. Petrology 47*, p. 610-629.
Bloom, A.L. (1970). Paludal stratigraphy of Truk Ponape and
   Kusiae, eastern Caroline Islands. *Geol. Soc. Am. Bull. 81*,
   p. 1895-1904.
Brunn, P., (1962). Sea-level rise as a cause of shore erosion.
   Proc. Am. Soc. Civ. Eng., *Jour. Waterways and Harbors Div.
   88*, p. 117-130.
Curray, J.R. (1965). Late Quaternary history, continental
   shelves of the United States. *In* "The Quaternary of the
   United States"(H.E. Wright and D.G. Frey, eds.), p. 723-
   735. Princeton Univ. Press, New Jersey.
Daniels, R.B., Gamble, E.E., Wheeler, W.H., and Holzhey, C.S.
   (1977). The Arapahoe ridge - A Pleistocene storm beach.
   Southeastern Geology 18, no. 4, p. 231-247.
Emery, K.O. and Milliman, J.C. (1971). Quaternary sediments of
   the Atlantic continental shelf of the United States. *In*
   "Colloque sur l'évolution des côtes et des plateformes
   continentales dans leur relation mutuelle pendant le
   Quaternaire" (A. Guilcher, ed.). Quaternia, v. 12, p. 3-18.
Emery, K.O. and Uchupi, E. (1972). Western North Atlantic Ocean:
   topography, rocks, structure, water, life, and sediments.
   Am. Assoc. Petroleum, Geol. Memoir 17, 532 p.
Field, M.E. and Duane, D.B. (1976). Post-Pleistocene history of
   the United States inner continental shelf: Significance to
   origin of barrier islands. *Geol. Soc. Amer. Bull. 87*,
   p. 691-702.
Fisher, J.J. (1962). Geomorphic expression of former inlets
   along the Outer Banks of North Carolina. Unpubl. M.S. thesis, Univ. of No. Carolina, Chapel Hill, 120 p.
Fisher, J.J. (1968). Barrier island formation, discussion.
   Geol. Soc. America Bull. 79, p. 1421-1426.
Godfrey, P.J. (1970). Oceanic overwash and its ecological implications on the Outer Banks of North Carolina: Office
   of Natural Science Studies, Annual Rept. National Park
   Service, Washington, D.C., 37 p.
Hayes, M.O. (1979). Barrier island morphology as a function of
   tidal and wave regime. *In* this volume.

Hoyt, J.H. (1967). Barrier island formation. *Geol. Soc. Amer. Bull. 78*, p. 1125-1136.

Kraft, J.C. (1971). Sedimentary facies patterns and geologic history of a Holocene marine transgression. *Geol. Soc. America Bull. 82*, p. 2131-2158.

Kraft, J.C. (1976). Radiocarbon dates in the Delaware coastal zone (Eastern Atlantic Coast of North America): Delaware Sea Grant Technical Report, DEL-SG-19-76, 20 p.

Milliman, J.C. and Emery, K.O. (1968). Sea levels during the past 35,000 years. *Science 162*, p. 1121-1123.

Nixon, R.B. and Pilkey, O.H. (1976). Reconnaissance geology of the Cape Lookout Quadrangle, North Carolina. U.S. Geol. Survey Prof. Paper 859, 45 p.

Moslow, T.F. and Heron, S.D. (1978). Relict inlets: preservation and occurrence in the Holocene stratigraphy of Southern Core Banks, North Carolina. *Journ. Sed. Pet. 48*, no. 4, p. 1275-1286.

Otvos, E.G. (1970). Development and migration of barrier islands, northern Gulf of Mexico. *Geol. Soc. America Bull. 81*, no. 1, p. 241-246.

Pierce, J.W. and Colquhoun, D.J. (1970). Holocene evolution of a portion of the North Carolina coast. *Geol. Soc. America Bull. 81*, no. 12, p. 3697-3714.

Susman, K.R. and Heron, S.D., Jr. (1979). Evolution of a barrier island, Shackleford Banks, Carteret County, North Carolina. *Geol. Soc. America Bull. 90*, p. 205-215.

Swift, D.J.P. (1975). Barrier-island genesis: evidence from the central Atlantic shelf, eastern U.S.A. *Sedimentary Geology 14*, p. 1-43.

GEOMORPHOLOGY, WASHOVER HISTORY, AND INLET ZONATION:
CAPE LOOKOUT, N.C. TO BIRD ISLAND, N.C.

William J. Cleary
Paul E. Hosier

Program in Marine Sciences
University of North Carolina
Wilmington, North Carolina

*Seventeen islands and two mainland beaches comprise the
237 km shoreline between Cape Lookout and Bird Island, North
Carolina. The area, characterized by diverse physiography, has
been divided into four major geomorphic sections. Shackleford,
Bogue, Bear, and Brown Islands comprise the northernmost 75 km
of shoreline. These barriers display vegetated beach ridges,
massive dunes and few washovers. Wide shallow lagoons with
little tidal marsh vegetation back these islands. The 50 km
section to the southwest consists of Onslow and Topsail Beaches,
where washovers are more abundant; many have occurred prior to
1938. Inlet zone maps indicate that inlets are more active in
this section than the previous area. These maps are based on
the location of marsh islands found in the vegetated lagoons,
and historic maps and charts. The 60 km section from Lea
Island to Fort Fisher-South includes six islands and the Caro-
lina-Kure Beach mainland. Inlets have been particularly active
within this area during the past 150 years, accounting for over
68% of the surface sediments. This area has been impacted by
washovers chronically during the past 40 years and vegetation
patterns suggest occurrences which predate 1900. Washover-
related physiography includes a number of identifiable shoreline features which suggest a cyclic pattern of washover followed by recovery. Both physiographic and vegetational recovery
patterns differ, depending upon the grain size distribution of
the washover sediments. It is postulated that changes in the
barrier islands during the past 3500 years have been most significant in the Lea Island to Fort Fisher-South section, creating absent, scattered, and single dune ridge morphology along
much of the shoreline. West of Cape Fear to Bird Island, the*

remaining 52 km consists of the Caswell-Yaupon-Long Beach main-land and Holden, Ocean Isle, Sunset, and Bird Islands. Suscep-tibility to washover within the region ranges from high to low, while inlets have had a significant impact on the area. The interrelationships among washover potential, vegetation cover, and dune morphology have been used to develop a generalized model of the shoreline features and processes of the Cape Look-out to Bird Island section of North Carolina.

INTRODUCTION

The North Carolina coastline consists of a sequence of large capes and associated shoals, barrier islands, spits, and occasional headland areas (Duane and Field, 1976). A natural division of the North Carolina coastline occurs near Cape Lookout (Fig. 1). North of this Cape, the islands are separated from the mainland by relatively wide open water lagoons and sounds, contrasting with the south where the sounds that back the barrier islands are narrow and nearly filled with marsh.

It has generally been accepted that the North Carolina coastline as a whole is undergoing a gradual transgression that effectively translates the barrier complex landward through a combination of processes: inlet dynamics, aeolian transport, and oceanic overwash. In the past, most studies involving North Carolina shorelines have been regional in scope, providing little detail, or have centered on barrier islands north of Cape Lookout, principally those within the Cape Hatteras National Seashore. Those studies which are general in nature include El-Ashry and Wanless (1968), and Langfelder et al. (1968). Other regional investigations include the inlet and geomorphic studies of Fisher (1962), Langfelder et al. (1974), Nummedal et al. (1977), and Baker (1977).

More detailed studies of those islands in the proximity of Cape Hatteras include those of Pierce (1969), U.S. Army Corps of Engineers (1964), Athearn and Ronne (1963), Shideler (1973), and Pierce and Colquhoun (1970). More recent studies emphasized the role that oceanic overwash plays in the development of these islands during landward migration. Studies by Godfrey and Godfrey (1973), Dolan (1973), and Schwartz (1975) have maintained that the islands migrate landward in response to a rise in sea level via overwash which normally accompanies strong tropical and extratropical storms. By this process, the islands develop an equilibrium profile and ultimately exhibit a relatively low, flat profile such as Core Banks and Portsmouth Island. Vallianos (personal communication, 1975) discounted the importance of oceanic overwash as a viable process for maintaining the integrity of the islands. His observations, however, were made

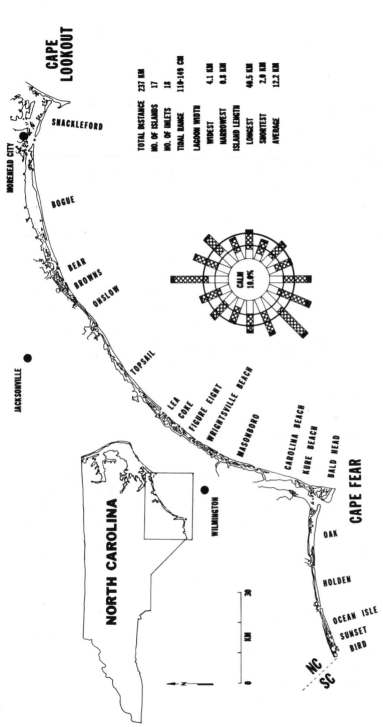

Fig. 1. The study area is located in SE N.C. extending from Cape Lookout to Bird I., N.C. The area includes 17 barrier islands & 2 mainland sections:Carolina Beach & Oak I. Percentage freq. of wind direction & speed (1948-1967) is represented by rose diagram. Wind speed (knots) represented on bar graph are: 1-6(white); 7-16(cross-hatched); & 17-27(black). Speeds greater than 27 knots comprise 0.2% of the total for the 20-yr. period. Concentric circles = a frequency increase of 2%.

in areas north of Cape Lookout, where sand supply is large in comparison to the southeastern barrier islands near Cape Fear.

In the vicinity of Cape Lookout, deep cores on Shackleford Island have shown that much of the underlying sedimentary sequence is composed of inlet fill (Sussman and Heron, in press). Data from a similar set of deep cores have shown that most of Core Banks, northeast of Cape Lookout, is underlain by a transgressive sequence of coarse washover sediments resting atop finer back-barrier sediments (Moslow and Heron, 1978). Other detailed studies were focused on engineering aspects of the coastal islands. Investigations by the U.S. Army Corps of Engineers involved analysis of a 19 mile section of the shoreline between Wrightsville and Carolina Beaches, and a 25-mile section of the shoreline south of Cape Fear. These studies determined the littoral transport rates and the impact of man-made structures on shore processes (Magnuson 1965; Vallianos 1970, 1975; Jarrett 1977; U.S. Army Corps of Engineers, 1973).

A more exhaustive survey shows that the barrier island complexes that form the coast south of Cape Lookout (Fig. 1) have received little detailed study. The present investigation was designed as a reconnaissance study to: (1) assess the role of inlet sedimentation, (2) ascertain the importance of oceanic overwash, (3) develop models for physiographic and vegetation succession patterns following washover, and (4) devise a shoreline classification that reflects washover potential for the southeastern section of North Carolina.

STUDY AREA CHARACTERISTICS

The seventeen barrier islands (Table 1) that comprise the 237 km coastline between Cape Lookout and the South Carolina line have a wide variety of physiographic forms, ranging from overwash-dominated narrow barriers to wide barriers with massive dunes and no washovers. The longest of these islands is Bogue Banks, 40 km long; Lea and Bird Islands are the shortest in the chain, approximately 2 km long. Lagoons are widest (4.1 km) in the north, behind Bogue Banks, and generally decrease and finally disappear where the islands have overridden the mainland section of Cape Fear near Carolina and Yaupon Beaches (Oak Island). The northern lagoons, principally Bogue Sound, are largely shallow, open, and free of vegetation. By contrast, tidal marshes generally have infilled the lagoons in the remaining portions of the study area. Fresh-water marshes occur only along the central portion of the Oak Island section of Long Beach, west of Cape Fear. Elevations on the islands range from less than 4 meters on Masonboro Island to more than 10 m above MSL on Bear Island (Fig. 2).

Table I. Barrier island characteristics. Cape Lookout to Bird Island, N.C. Mean widths were calculated from transects established at 1 km intervals on aerial photographs.

Island	Width m	Length km	Mean lagoon width m
Shackleford	494	14.78	5273
Bogue	618	40.49	2802
Bear	555	5.57	2587
Brown	323	5.73	1376
Onslow	201	12.13	888
Topsail	287	36.25	1650
Lea	220	2.80	2562
Coke	134	4.57	2537
Figure Eight	201	7.01	1754
Wrightsville	220	7.53	1834
Masonboro	153	13.07	1672
Carolina	140	4.96	618
Fort Fisher-South	214	4.75	*
Bald Head	1000	8.84	*
Oak Island	238	20.67	415
Holden	250	13.17	530
Ocean Isle	353	9.15	796
Sunset	378	3.32	1525
Bird	335	2.09	2105

* No lagoon present behind beach.

These low mesotidal barrier shorelines (Hayes, this volume), have a mean tidal range that increases from 110 cm in the north (Atlantic Beach) to 135 cm in the south (Oak Island). The direction of wind approach fluctuates annually. During the spring and summer the winds are from the south and southwest, while during the winter, north and northeast winds prevail. The section of the study area south and west of Capes Lookout and Fear are more protected from northeast storms than the area north of Cape Fear. All sections are highly vulnerable to hurricanes approaching from the south and east (Carney and Hardy, 1967).

Inlets vary from wide, deep, stabilized and maintained inlets such as Beaufort Inlet, separating Bogue and Shackleford Banks, to narrow, shallow, shifting inlets such as Mason's and Rich's Inlets, separating Wrightsville and Figure Eight Islands, and Figure Eight and Coke Islands. Human impacts on the islands vary from extensive development on portions of Wrightsville Beach, Bogue Banks, and Topsail Island to uninhabited islands such as Bird, Lea and Bear Islands.

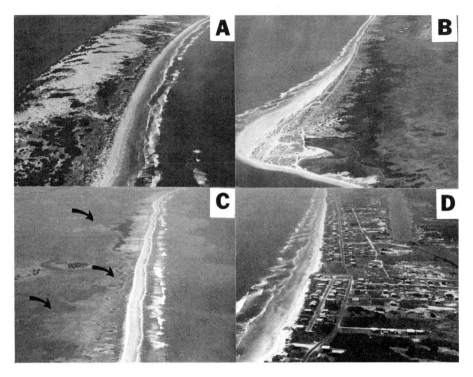

Fig. 2. (A) This island is typical of Area I islands with massive dunes. Bear Island exhibits little evidence of over-wash. (B) Topsail Island. Certain areas such as the north end of Topsail Island in Area II have been influenced by overwash prior to 1938, the date of the earliest available aerial photo-graphy. The shrub thicket marks the edge of late 19th century washover fans and terraces (arrow). (C) Masonboro Island. Ma-sonboro Island, located in the southern part of Area III, is typical of the narrow, steep and coarse sandy barrier island sections that occur within the proximity of Cape Fear. These islands have been repeatedly influenced by overwash. Note abundant washover fans (arrows). (D) Holden Beach. Holden Beach is one of the islands characteristic of Area IV. Islands are alternately high with extensive, massive dunes (foreground); and wide, flat and washover dominated (background).

INLET SEDIMENTATION

Locations of former inlets or historic inlet zones were de-termined from two basic lines of data. An analysis of sequen-tial aerial photographs (1938-1977) provided information re-garding the contemporary history. This data set provided information for those areas influenced by cutting/infilling and consequent spit elongation during this forty-year period. These data also served to identify areas prone to inlet formation during storms. Aerial photographs also provided a data base for assessing the importance of inlet sedimentation by delineating the geomorphic expressions of former inlets that pre-date 1938. These physiographic features included vegetated flood-tidal deltas (Fig. 5) (Godfrey and Godfrey, 1974). In the narrow, partially marsh-filled lagoons of the study area are elongate islands indicative of former inlet locations.

These islands, which commonly parallel the main tidal channels and the general barrier island shoreline, are usually vegetated and in some instances surrounded by tidal marsh (Fig. 6A). Development of these features occurs at a point where sands from the flood-tidal shield overtop the adjacent tidal marsh. The major building process is related to the proximity of the flood-tidal sand sheet and increased wave swash in the vicinity of the inlet gorge or main channel. Storms are likely to be major factors in their formation and alteration (Cleary et al., submitted). Continued migration of the inlet leads to development of additional islands and subsequent preservation of earlier formed features as sand supply and wave approach is reduced (Figs. 6B and 7).

Fig. 8 illustrates the topography and distribution of vegetation across a typical marsh island. Elevations commonly range from less than 40 cm to as much as 300 cm above mean high water. Shallow cores show that the islands rest on relatively thick organic-rich salt marsh peat. Vegetation across these features occurs within well-defined zones, with the non-salt tolerant species of grasses and shrubs restricted to crestal elevations. Older islands and those with substantial relief usually have a more mature vegetation cover. The effects of a sea-level rise are observed in the *Juncus* marsh, where drowned stump fields of decay-resistant cedar are found.

A second data set used to locate former inlet zones was derived from inspection of historic coastal maps and charts. Womble's 1738 map appears to be the first and most accurate in terms of delineating the study area. Base maps for sections of the shoreline were produced from 1974 aerial photographs. Chart sets were then enlarged to the common scale. Locations of inlets with historical significance were checked with records and documentaries.

Figure 3 illustrates the history of New Topsail Inlet, and is similar to the type of map which was produced for sections of the shoreline. Inlet zonations were then delineated, based on these compilations (Fig. 9).

As a method of providing a check on the reliability of the geomorphic and historic data, a program involving the recovery

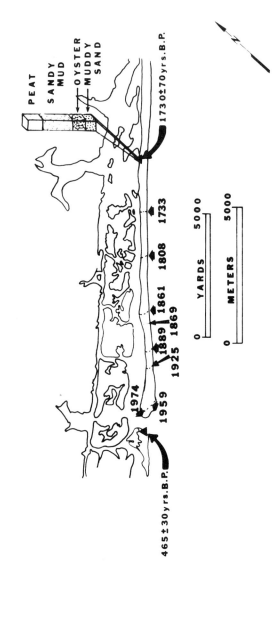

Fig. 3. NEW TOPSAIL INLET. Generalized history of New Topsail Inlet. Former inlet locations were determined using data derived from historic charts, maps, aerial photographs, and geomorphic features.

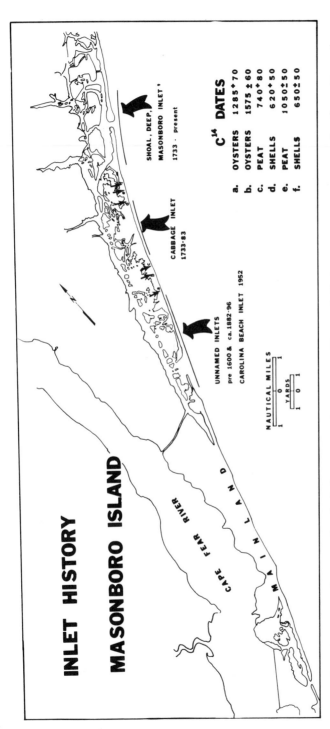

Fig. 4. Masonboro Island inlet zonation map. Three inlet zones were recognized on Masonboro Island. Shallow 3 m cores substantiated the lack of inlet sediment and presence of old lagoonal and tidal marsh sediments. The oldest datable material was recovered from the narrowest sections of the island.

Fig. 5. Relict flood tidal delta on Core Banks. Similar geomorphic evidence in the form of vegetated flood-tidal deltas in the wide lagoons provide supporting evidence for the location of former inlets within the study area (see Cabbage Inlet, Fig. 4).

of shallow cores (<3 m) was undertaken. Figure 4 shows the location of six radiocarbon-dated samples recovered from beneath the grassland surface and from outcropping lagoonal sequences on the foreshore. From these dates, which range from 620 ± 50 to 1575 ± 60 yrs BP, it is apparent that certain sections of Masonboro Island and the adjacent lagoon have not been influenced by inlet sedimentation for a relatively long time when compared to adjacent sections. The location and age of these samples also lends support to the reliability of the geomorphic and historic data.

A synthesis of this information was used to construct an inlet zonation map for the study area (Fig. 9). Geomorphic evidence indicates that a total of 42% of the study area has been influenced by inlets, and historic evidence indicates 45%. The combined total (Table 2) suggests that 56% of the coastline is underlain by inlet fill. These estimates are conservative.

An earlier study of the Outer Banks (Bogue to Currituck Banks) has shown that historic inlets accounted for 15% of the underlying sediments while pre-historic inlets occurred over 45% of the shoreline (Fisher 1962). Data from the northwest

TABLE II. *Percentage Historic and Geomorphic relict inlet zones in the four physiographic areas, Cape Lookout to Bird Island, North Carolina.*

Area	Geomorphic (%)	Historic (%)	Total (%)
1	16.03	37.16	42.31
2	58.33	56.48	68.06
3	57.56	59.66	73.53
4	47.24	34.65	48.03
Total (%)	42.45	45.88	56.47

Gulf of Mexico have shown that tidal-inlet sediments comprise up to 35% of the sedimentary sequence (Shepard & Moore, 1955).

Historically, inlets play a lesser role in island and lagoon sedimentation in areas 1 and II (Table 2, Fig. 9). Geomorphic evidence is least apparent in area 1, which is characterized by wide open and shallow lagoons and wide east-west oriented islands with a large sand supply. South of area I to Cape Fear, geomorphic data are much more evident.

In Area III, inlets have occurred along 74% of the shoreline. The section south of Topsail Island is backed by narrow (<2 km) marsh-filled lagoons. These islands range in length from 2 to 13 km and are among the narrowest in this area. The shoreline section south of area 1 is apparently transitional in nature between mesotidal and microtidal barrier island shorelines characterized by Hayes (this volume).

WASHOVER PHYSIOGRAPHY AND HISTORY

Aerial photographs dating from 1938-1977 were analyzed for shoreline physiography related to historic washovers within the study area. Vegetation destruction and foredune removal on aerial photographs were used as indicators of washover. With most photography at a scale of 1:20000, only washovers with dimensions of 50 by 50 meters which breached a portion of the foredunes were recognizable. Overwash was found to be an uncommon process based on inspection of nine sets of photographs available for the islands. Significant periods of washovers occurred during the periods of 1954-55 and 1962. These dates correspond with Hurricanes Hazel, Connie, Diane and Ione (1954 and 1955) and the Ash Wednesday Storm (1962). There are physiographic and vegetative characteristics which indicate that some areas have received washovers in the past, prior to aerial

Fig. 6. (A)Linear marsh islands (arrows) located within partially marsh-filled lagoons provide supporting evidence for the
location of former inlets. Islands with high relief undergo succession & may be dominated by arborescent vegetation. (B)Marsh
is. formation. The islands typically form at a point where sands
from the flood-tidal delta shield overtop the adjacent marsh.

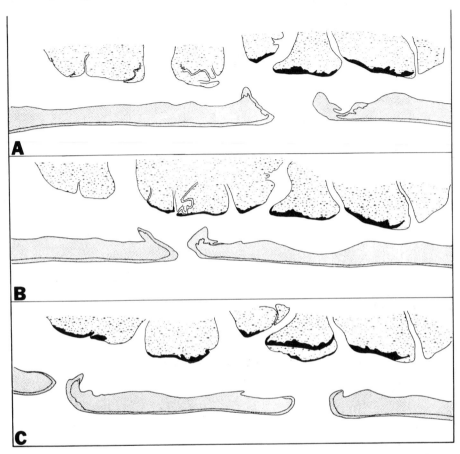

Fig. 7. MARSH ISLAND FORMATION. Idealized marsh island formation. (A) Linear islands form landward of the gorge of an inlet where flood tidal-delta shield sands overtop the marsh. (B) Migration of the inlet results in a chain of preserved marsh islands soundward of the inlet migration path. (C) A new inlet episode results in multiple marsh islands in the marsh-filled lagoon.

photographic coverage. These will be discussed later. With the exception of these events, however, major washovers are not evident.

Based on a study of Masonboro Island (Hosier and Cleary, 1977), four physiographic shoreline types related to washover events were recognized. The physiographic types include: (a) *Intact dunes.* These sections exhibit no recent washovers. The foredunes are generally intact and continuous, although sometimes scarped (Fig. 10A). Dune vegetation cover is very dense,

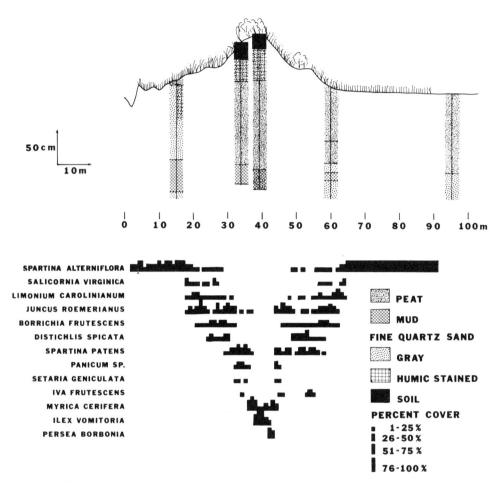

Fig. 8.   Marsh Island Transect.   The islands are generally narrow, with low relief ranging from 0.4 m to 2.5 m (MHW). Shallow cores indicate that the island is resting on salt marsh peat. Vegetation is zoned with less salt tolerant herbaceous and arborescent vegetation restricted to the crest of the island.

and shrub thickets or maritime forest may be present. Shorelines which have received no washover impacts are categorized in this group. (b) *Isolated washover fans* (Fig. 10B). This physiographic type generally occurs in zones along the beach-dune system. The dunes are breached and recent unvegetated or poorly vegetated washover fans are evident. (c) *Washover terraces*. Where washovers have completely erased the foredunes (Fig. 10C), individual fans coalesce and form washover terraces. This physiographic-vegetational type is evident for 5 to 15

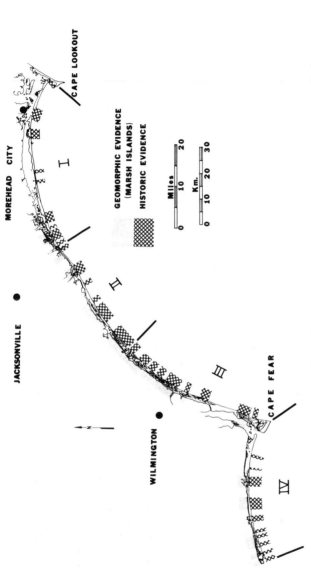

Fig. 9. INLET ZONATION. Spatial distribution of historic and geomorphic inlet zones. Inlet zones are especially evident in the southeastern section of Area II and the northeastern section of Area III. The wide, high islands of Area I exhibit low inlet frequency, both historic and geomorphic.

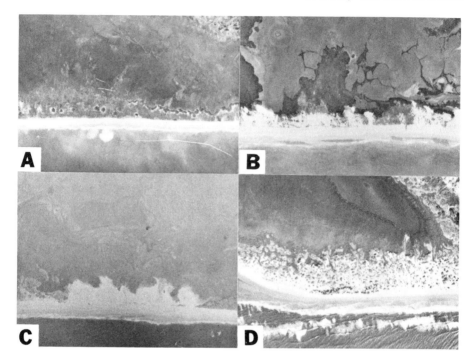

Fig. 10.   Physiographic types recognized on aerial photographs. (A) Intact dune ridge. Crenate patterns of shrubs indicate former washover fan margins. (B) Isolated washover fans.
Individual fans penetrate the foredune ridge, often extending
across the island. (C) Washover terrace. The terrace is formed
by coalesced washover fans. The foredune ridge is obliterated.
(D) Dune recovery. Recovery is initiated by development of
scattered "haystack" dunes above the beach berm on the washover
fan surface.

years following the washover impact as estimated by the density
of vegetation cover. On Masonboro Island, for example, washover
fans are clearly evident for 12 years following overwash (Fig.
12A). Vegetation is slow to recover on the washover fan. On
islands where vegetation recovery is complete, the outline of
the fan is often clearly evident, indicating a washover event
(Fig. 12B). (d) *Foredune recovery*. The fourth physiographic
type recognized is dominated by small dune fields or ridges
suggesting dune recovery following overwash (Fig. 10D). The
recovery dunes are usually discontinuous above the berm line
with adjacent intact dune ridges.
      Each of these types was mapped on available photography for
the length of the study area. Figure 11 shows the temporal and

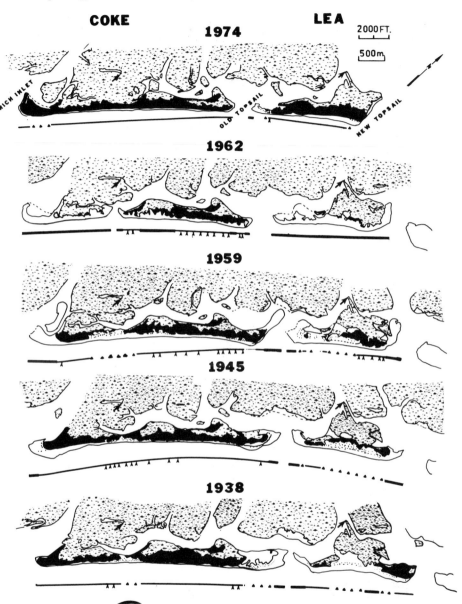

Fig. 11. Aerial photograph compilation of Lea & Coke I. (1938–1974) based on the physiographic types identifiable on aerial photos (Fig. 10). ___ = intact dune ridge; $\overline{\Lambda\Lambda}$ = isolated washover fans; ▬ = washover terraces; & ▲ = washover recovery zones. The historical trends of the islands, including inlet migration & beach-dune morphology are evident. Evidence of the major washover episodes of 1954 and 1962 are present on these islands.

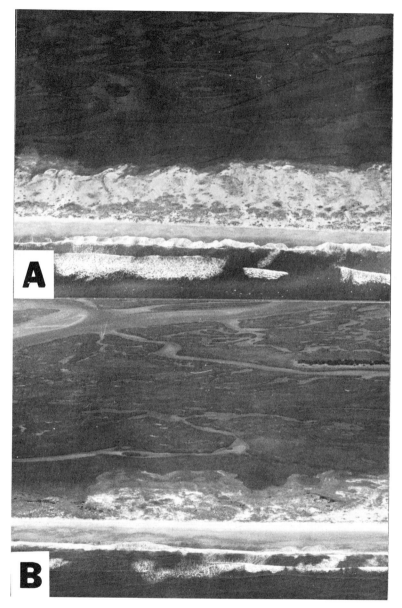

Fig. 12. (A) Masonboro I. (1974). Washover fans occurred in
'54 and '62. Coarse shell material produced lag surfaces retard-
ing redevelopment of dunes & vegetation cover on the surface of
the washover terrace. (B) Coke I. (1974). Washover fans occurred
in '54 & '62. Fine sandy substrates are remobilized into a well-
vegetated dune field & foredune ridge. With time, shrubs redevelop
beginning along the distal margin of the fan (cf. Fig. 2B & 11A).

spatial distribution of these physiographic shoreline features
on Lea and Coke Islands. All physiographic types are evident
within the area during the period 1938-1974. Isolated fans
(Fig. 10B) and washover terraces (Fig. 10C) are dominant in
1959 and 1962 photography. Stable dunes predominate in 1974
photography, indicating a period of relative stability in recent years. These maps thus provide a data base for determining the washover history of specific areas. Analysis of these
maps pinpoint stable beaches as well as zones of severe or chronic washover impact. Further, the progress of recovery can be
qualitatively evaluated from the aerial photography using
vegetation density as a marker.

Topsail Beach (Fig. 2B) and Masonboro Island (Fig. 2C) are
examples of islands which exhibit an earlier generation of morphological features associated with severe overwash. These
areas are characteristic of physiographic type B; dunes have
redeveloped following the severe washover. The crenate pattern
of shrub thickets indicates the landward margin of washover
fans which predate 1900. On Topsail Beach, 1954 and 1962 washover fans extended to the edge of the shrub thicket but did not
penetrate it. Observations indicate that all or portions of the
shrub thicket may be killed by large washover events; however,
the mode of action is not understood. Salt water flooding, partial sand burial, or uprooting of the thicket could account for
the shrub destruction.

Initial studies on Core Banks (Godfrey and Godfrey, 1976)
and Masonboro Island (Hosier and Cleary, 1977) suggested that
washovers along the North Carolina coast were predominately
coarse-grained. Vegetation recovery on the fan surfaces of these
islands which received severe and chronic washover impacts was
very slow. Figure 12A indicates the extent of vegetation recovery on the fan surfaces of Masonboro Island 12 years after
the Ash Wednesday Storm (1962). Dune ridge redevelopment is
similarly retarded; the washover fan exhibits only low discontinuous "haystack" foredunes. A model has been produced that
depicts the cycle of washover and recovery on coarse-grained
washover areas within the study area (Fig. 14). The narrow,
poorly developed dune ridge is easily breached by the surge
accompanying hurricanes or by wave attack from northeast storms.
Coarse-grained beach and foreshore sands and shell, as well as
dune sands, are washed across the island and deposited as fans.
Coarse shell rapidly armors the surface as deflation occurs,
and nearly flat, featureless areas behind the beach berm are
produced. Foredune redevelopment is similarly retarded and the
area remains susceptible to washover. In fact, the Ash Wednesday Storm (1962) enlarged many washover fans produced by hurricanes in 1954-55.

In contrast, a number of islands within the study area
which are overwashed by predominately fine-grained sands

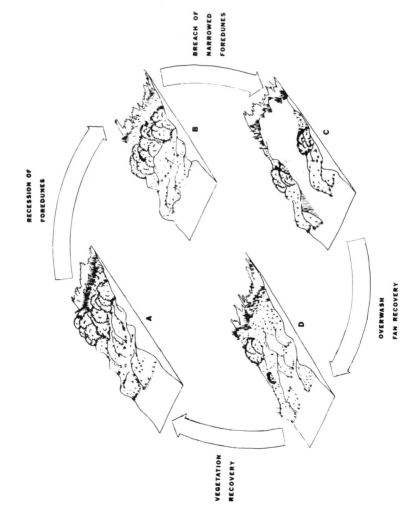

Fig. 13. Fine-grained washover model. Recession of foredunes (B) allows breach of ridge with subsequent washover fan formation. Herbaceous and arborescent vegetation are removed during washover fan formation (C). Remobilization of sands on the fan surface and foreshore areas (D) create a dune field and re-established foredune ridge (A). The extent of recovery makes the area less susceptible to subsequent washovers.

RECESSION OF FOREDUNES

BREACH OF NARROWED FOREDUNES

A

B

C

D

OVERWASH FAN RECOVERY

VEGETATION RECOVERY

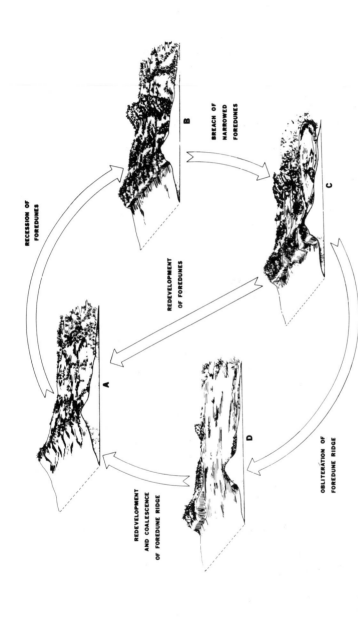

Fig. 14. Coarse-grained washover model. Scarped dune ridge (B) continually recedes with impinge-
ment of storms in area. A breach may occur in a section of narrowed dunes (C) & resultant washover
sediments are carried thru the dune field onto the back-barrier, elevating tidal & subtidal environments. Continued washover across the low dune ridge gradually eliminates intervening dunes, resulting in terrace-like features with low relief. Following a period of quiescent conditions, redevelopment & coalescence of foredune ridge occurs (A). It is possible that individual washover
fans (C) could quickly recover and develop a coalesced dune ridge (A).

257

exhibit a different pattern of recovery. Coke Island is an example (Fig. 12B). Coke Island was overwashed in 1954 and again
in 1962; however, the fine-grained sands of the fan surface were
quickly remobilized and a dune field developed (Fig. 13). Revegetation of the area is primarily via establishment of sea
oats as the dunes reform. Availability of abundant sand in the
foredune area has resulted in reestablishment of a continuous
high foredune ridge. The combination of the foredune ridge and
an extensive dune field on the washover surface has created a
shoreline on Coke Island which is less susceptible to washovers
than on Masonboro Island.

Figure 15 summarizes the washover history for the area from
Cape Lookout to Bird Island. The map is derived from aerial
photographic compilation (Fig. 11). The scale for the washover
history is based upon frequency of washover events along the
shoreline sections. Three major episodes of washover form the
basis of the classification: pre-1938 (based on vegetation patterns), 1954-55, and 1962. Those areas exhibiting washover during all three periods are classified as severe; two washover
periods indicate moderate, and one washover episode is classified as occasional. The shorelines exhibiting no washovers were
classified none.

Area I has received the least washover impact. The extensive shoreline dominated by multiple dune ridges on Bogue Banks
reduces the washover section in area I. Areas II and III have
significant washover sections, while area IV exhibits occasional
and moderate impacts (Fig. 16). The Ash Wednesday Storm (1962)
had little effect on this section of the North Carolina coast.

Generally, the areas receiving impacts during the past 80
years are highly susceptible to subsequent washover. Of the total shoreline impacted by washover, 83.7% has been influenced
by 2 or more events. These areas are predominately sections
which exhibit single dune ridges (Fig. 20). The pattern suggests that in response to present transgressive conditions,
barrier islands undergo a morphological evolution whereby the
islands recede until washovers become an important mechanism
for landward sediment transfer. Once this influence begins,
washovers dominate the physical processes affecting the shoreline.

In order to develop an empirical estimate of washover
susceptibility for the barrier islands in the study area, transects drawn perpendicular to the beach were established at 1 km
intervals on 1974 aerial photographs. Total island width, herbaceous vegetation width, arborescent vegetation width, and
erosion rate (Langfelder et al. 1968) were measured along the
transect. These data were standardized and summed for each
transect.

The standardized sums are plotted for each point in Figure
17. These standardized values were correlated with washover

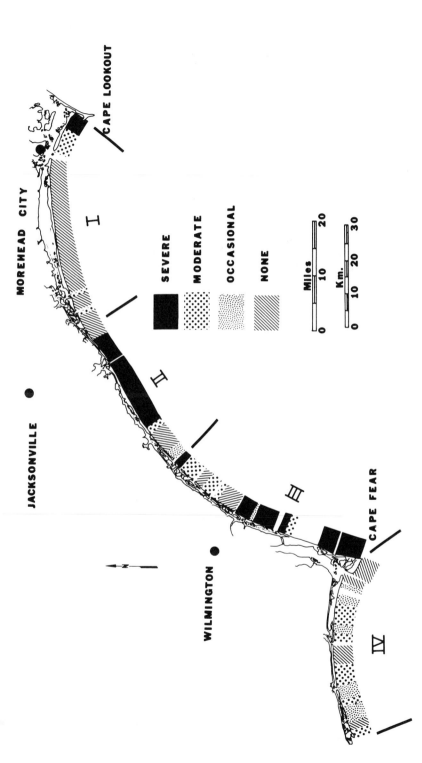

Fig. 15. WASHOVER HISTORY SUMMARY MAP. The washover history is based on frequency of major over-wash episodes. Areas of severe washovers have been impacted at least 3 times in the last 80 yr. The moderate category exhibits 2 washover episodes & the occasional category only one during the same period. Areas II and III have received the greatest impact.

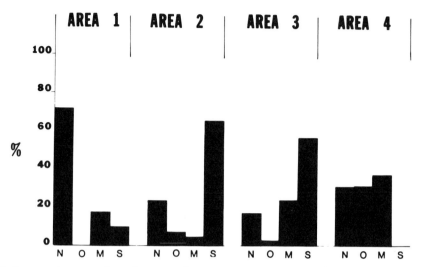

N=None; O=Occasional; M=Moderate; S=Severe washover history

Fig. 16. WASHOVER HISTORY. Washover history for the major subdivisions of the study area. From north to south, the percentage of moderate and severe washover influences increases, while the island sections experiencing no washovers decreases.

history. An inverse correlation was established (r = -0.9597). Higher standardized sums were significantly correlated within frequent washover history. On this basis, areas of severe, moderate, and low susceptibility were assigned to the standardized sums for each susceptibility level. General trends of the figure show that islands from Shackleford to Brown Islands, and from Bald Head to Bird Island are generally less susceptible to washover than are islands from Onslow to Fort Fisher-South. Major sections of Coke, Figure-Eight, Masonboro, Carolina Beach, and Fort Fisher-South are highly susceptible to washover based on the empirical parameters included in the standardized sum. We feel that this is a first step in developing a predictive tool to determine areas prone to washover along the southeastern North Carolina coast.

DUNE MORPHOLOGY

The extent and integrity of the dune system varies along the 237 km shoreline. The nature of the dune system was analyzed from aerial photographs. The 1974 photograph set was thought to be representative of the nature and morphology of

the dune system during a period of relative stability. The last
major event that affected the coastal area was the Ash Wednesday Storm (1962).

Four distinct types of dune morphology were recognized from
inspections of the imagery (Fig. 18): (a) those sections where
dunes are absent or scattered (Fig. 18A); (b) areas where a
single dune ridge is present (Fig. 18B);(c) shorelines where
multiple dune ridges are prominent (Fig. 18C); and (d) those
sections where massive dunes occur (Fig. 18D).

In area I (Fig. 19) large multiple dune ridges with elevations in excess of 12 m or massive dune systems occur over 90%
of the shoreline. The combined total of these two categories
decreases markedly southward toward Cape Fear. In this section
of the study area, the shoreline is characterized by scattered
dunes or sections with no dunes. Approximately 80% of area III
has a single dune ridge or a redeveloping foredune ridge (scattered). This pattern is typical of areas with moderate to
severe histories of overwash.

South of Cape Fear, the dune system is composed essentially
of single scarped dune ridges or massive dune fields. It is
interesting to compare the nature of the dune system in areas
I and IV, both of which are east-west trending shoreline sections. Unlike area I, the section west of Cape Fear generally
has a more severe history of washover, as reflected in the integrity of the dune system. Both sections generally escape the
severe damage that occurs in area III due to northeasters. Part
of the difference in the dune systems reflects an inherited
morphology from a stillstand 3500 years ago when the majority
of area I began to prograde, the results of which are a series
of well vegetated beach ridges (Heron and Hine, personal communication).

Other factors to consider include amount of sediment in the
longshore transport system, shoreline rate, and frequency and
migration characteristics of tidal inlets (Fig. 9).Tidal deltas
would act as sediment sinks or storage areas. Data indicate
that while a more significant volume of sand is transported in
the area III system than in other shoreline segments, it has
one of the higher erosion rates. Complicating this picture of
sand supply is the impact of major hurricanes. Inspection of a
plot of the major nineteenth and twentieth century storms shows
that areas III and IV are more frequently impacted and suffer
more damage than areas southwest of Cape Lookout. Still-water
surge heights associated with a twenty-five year storm increases an average of 2 feet (60 cm) from Cape Lookout to Bird
Island (7.6 - 9.7 ft MSL). The relatively low and narrow dune
systems that occur along much of area IV are then more likely
to be overtopped when compared to other areas.

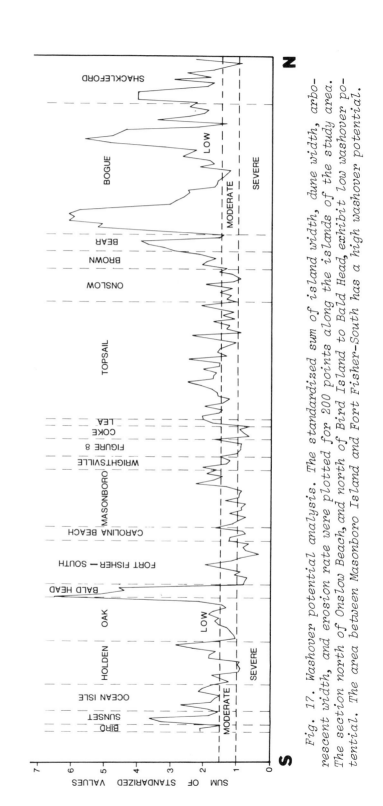

Fig. 17. Washover potential analysis. The standardized sum of island width, dune width, arborescent width, and erosion rate were plotted for 200 points along the islands of the study area. The section north of Onslow Beach, and north of Bird Island to Bald Head, exhibit low washover potential. The area between Masonboro Island and Fort Fisher-South has a high washover potential.

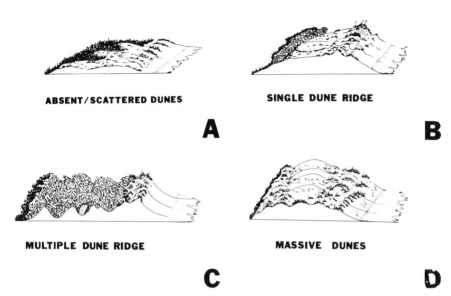

ABSENT/SCATTERED DUNES   **SINGLE DUNE RIDGE**

A   # B

MULTIPLE DUNE RIDGE   **MASSIVE DUNES**

C   # D

Fig. 18. Beach-dune morphology types recognized in the Cape Lookout to Bird Island portion of the N.C. coast. (A) Absent / scattered dunes. Island sections which have recently received washovers, or chronic washover areas, are included. (B) Single dune ridge. Certain island portions also have received significant washover in the past, but foredune ridges have recovered. (C) Multiple dune ridges. Segments of islands which have had a history of progradation exhibit well developed dune ridges. (D) Massive dune ridges. Areas with extensive mobile dunes and/or large sand volumes exhibit massive dune morphology.

Because of the frequency of hurricanes and higher storm surges associated with area IV, the integrity of the dune system is less than area I.

A similar correlation between frequency of hurricanes, storm surge levels, and nature of the dune system can be made for area III, a northeast-facing shoreline section. The 30 km spit and barrier island section south of Wrightsville Beach is in an area where the dune system is absent or poorly developed, or where a single scarped dune ridge occurs. Most recently, Jarrett (1977) has shown that major man-made shoreline alterations, including the artificial opening of Carolina Beach Inlet (1952) and the Masonboro Inlet jetty (1966), have dramatically altered the sediment budget and shore processes along the shoreline from Wrightsville Beach to Kure Beach.

Inspection of aerial photographs that predate the earliest

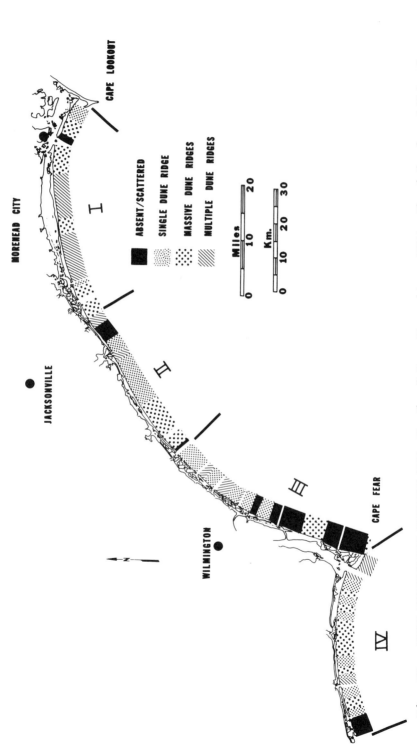

Fig. 19. DUNE MORPHOLOGY SUMMARY MAP. North of Cape Fear (Area III), the islands exhibit the greatest percentage of absent scattered and single dune ridges. Areas I and IV exhibit the most well developed dunes.

alteration (1949) shows that this section of the shoreline has relatively well developed dune ridges and fields. During the interim period (1952-66), the character of the islands and the dune system changed (Vallianos 1970, 1975; Hosier and Cleary, 1977). From 1954 to 1962, large numbers of moderate to severe hurricanes occurred within a short period of time. This storm climate may have produced an historical deficit of sand within this section, an area with an already altered sediment budget.

Synthesis of the information depicted in Figure 19 indicates that absent/scattered dunes account for 17.4% of the total shoreline section, single dune ridges 29.3%, massive dunes 31%, and multiple dunes 22.5% of the study area (Fig. 20). These differences are a reflection of an interplay of many factors including differences in wave energy regimes, island orientation, past storm histories, sediment sinks and storage in tidal inlet systems, and the occurrence of man-made structures.

ISLAND MORPHOLOGY: OVERVIEW

A basic and reasonable assumption is made that the entire area at one time was characterized by a series of barriers and spits that displayed well developed systems of beach ridges, such as those on Bogue Banks and Bald Head Island. Assuming this premise to be correct, it is apparent that area I has been characterized by the least amount of change, areas II and III the most, and area IV an intermediate amount. The more widespread occurrence of beach ridge barriers in the past has been suggested by Fisher (1962), Brown (1976), and Hine (personal communication). The ridges that presently exist within the study area reflect the general progradation during the stillstand, 3500 years ago.

Inspection of Table 2 and Figure 9 indicates that inlets play a more dominant role in area III and a less important role in area I in determining the island morphology and sediment patterns. Questions arise as to why there is a significant difference in morphology and inlet history along this section. Although conjectural, it appears that a significant number of factors interact to produce the observed patterns. These factors include differences in tidal regimes, lagoon characteristics, storm approach and frequency, and wave energy regimes.

If we compare the two end members (area I and III) there are significant differences in dune morphology (Fig. 20) and washover history (Fig. 15), and the percentages of shoreline along which inlets have occurred (Fig. 9). The tidal range difference in these two areas is less than 20 cm; this in itself is unlikely to account for the observed patterns.

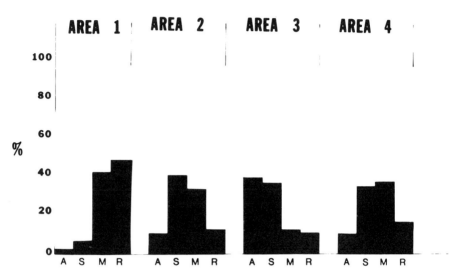

*Fig. 20. DUNE MORPHOLOGY.   Dune morphology for the major
subdivisions of the study area (S). From north to south absent/
scattered (A) and single dune ridge (S) morphology increases,
while multiple (R) and massive dunes (M) decrease in importance. This trend is related to the washover history (Fig. 15).*

Intuitively one might expect that the narrow, low barriers
that front the narrow lagoons in area III would show a higher
proportion of shoreline influenced by recent inlets than the
wide massive barriers fronting the wide, open water lagoons in
area I. Nummedal et al. (1977) have suggested that the variability of lagoon geometry ultimately controls morphology of the
tidal-inlet associated sand bodies (e.g., ebb-tidal deltas).
Aerial photographs indicate that the ebb-dominant inlets in the
southern section of North Carolina possess better developed and
larger ebb-tidal deltas (Hayes et al. 1973) when compared to
flood-dominant sections in the northern areas.

An inspection of historical charts and documents has shown
that within the past 125 years the shallow lagoons in the southern section have been significantly filled with fine-grained
materials and tidal marsh vegetation. The infilling of the southern lagoons and theoretically the concomitant changes in offshore storage in the ebb-tidal systems may have directly affected the sediment budget and indirectly the island morphology.

During this time of change in lagoon and inlet systems, a
minimum of 100 hurricanes (since 1804) and an unknown number of
winter storms have affected the North Carolina shoreline (U.S.
Army Corps of Engineers, 1973). The shoreline section near Cape

Fear has received the greatest impact from these events. The frequent overtopping of this shoreline section in comparison to other areas during this time period has been instrumental in contributing to the vulnerability of the shoreline to inlet formation and associated or subsequent washover events.

WASHOVER-VEGETATION CLASSIFICATION

Table III summarizes the relationships among dune, washover, and vegetation along the mainland and barrier island shorelines within the study area. The mainland areas of Carolina Beach and Oak Island are generally greater than 6 m in height and are dominated by arborescent vegetation. These shorelines have a low washover probability. The absent/scattered dunes and single dune ridge barrier island are the physiographic types most likely to be influenced by overwash. Coke, Lea, Figure-Eight, and Masonboro Islands are examples of this shoreline type. Sections which have not received overwash in the past 25 years may be dominated by arborescent vegetation. Massive dune fields and beach ridges are similar to mainland sections in respect to washover potential. The large sand buffer reduces the washover influence even in severe storms. These areas, such as Bogue, Bear, Browns, Bald Head Islands, and portions of Holden Beach, may receive small washover penetrations across one or two dune ridges, especially if the seaward dunes are narrow and/or discontinuous. On recently prograding beaches such as Sunset Beach and the northern half of Figure-Eight Island, washovers are uncommon. Herbaceous vegetation dominates these islands in contrast to the arborescent vegetation cover on the old beach ridge or massive dune islands.

TABLE III.   WASHOVER-VEGETATION BEACH CLASSIFICATION

	WASHOVER HISTORY	DOMINANT VEGETATION	WASHOVER POTENTIAL
MAINLAND	NO	ARBORESCENT	LOW
BARRIER ISLAND			
ABSENT/SCATTERED DUNES	YES	HERBACEOUS	HIGH
SINGLE DUNE RIDGE	YES	HERB./ARBOR.	HIGH
DUNE FIELD	NO	HERBACEOUS	LOW
BEACH RIDGE	NO	ARBORESCENT	LOW

CONCLUSIONS

The seventeen barrier islands and two mainland sections located between Cape Lookout and Bird Island, North Carolina,
possess a diverse inlet influence, washover history, and dune
morphology. Area I (Shackleford Bank to Brown Island) generally
possesses massive dunes and extensive dune ridges. Inlets have
been relatively unimportant and washover history is negligible
except for Shackleford Bank. Area II (Topsail and Onslow
Beaches) is characterized by washover-dominated topography,
often exhibiting several generations of washover fans. Extensive single-dune ridge morphology is common. Area III (Lea
Island to Fort Fisher-south) is highly washover dominated with
intensive inlet migration activity. Dune morphology is variable,
but sections of absent, scattered, or single-dune ridges predominate. Area IV (Bald Head to Bird Island) includes a diver-
sity of islands. Sections of Bird, Ocean Isle, and Holden are
low and flat, while other sections are characteristically high
with well developed dune fields.

Washover recovery in affected areas varies depending upon
the grain size distribution of the washover sediments. Both
physiography and vegetation redevelopment are slowed on coarse-
grained washover surfaces. Using total island width, arborescent vegetation width, herbaceous vegetation width, and erosion
rate, a correlation between these values and the washover history of an area has been produced. This method could be used to
develop a measure of the washover susceptibility for specific
sites within the study area.

It is suggested that many factors, including wave energy
climates, island orientation, severe storm history, tidal inlet
dynamics, and man's influence, interact to produce the variable
shoreline morphology of the study area.

ACKNOWLEDGMENTS

This report stems from several studies involving the barrier
islands of southeastern North Carolina. The research was sponsored in part by the North Carolina Board of Science and Technology (Grant #641); the Office of Sea Grant, NOAA, U.S. Department of Commerce, under Grant #04-6-158-44054; and the North
Carolina Department of Administration and the University of
North Carolina at Wilmington Marine Science Research Program.

Pamela Johnson, Jerry Williams, Glenn Wells, and Tim Griffin of the Earth Science Department, and Thomas Eaton of the Biology Department provided field assistance. Ed Helsing drafted most of the illustrations. A debt of gratitude is extended to Dr. Medha Kochhar for her patience and assistance during various stages of the study.

REFERENCES

Athearn, W.C. and Ronne, F.C. (1963). Shoreline changes at Cape Hatteras, an aerial photographic study of a 17 year period. Office of Naval Research, Washington, D.C., Naval Research Reviews, v. 6, p. 17-24.

Baker, S. (1977). The citizen's guide to North Carolina's shifting inlets. Sea Grant Publication UNC-SG-77-08, North Carolina State University, Raleigh, N.C., 32 p.

Brown, P.J. (1976). Variations in South Carolina coastal morphology. In "Terrigenous Clastic Depositional Environments: Some Modern Examples" (M.O. Hayes and T.W. Kana, eds.), p. II-101 - II-104. Tech. Rept. No. 11CRD, Dept. Geol., University of South Carolina.

Carney, C.B. and Hardy, A.V. (1967). North Carolina Hurricanes. Weather Bureau, ESSA, U.S. Dept. of Commerce, Washington, D.C., 40 p.

Cleary, W.J., Hosier, P.E., and Wells, G.R. (in press). Distribution and genesis of marsh islands in southeastern North Carolina lagoons. J. Sed. Petrol.

Dolan, R. (1973). Barrier Islands: Natural and Controlled. In "Coastal Geomorphology" (D.R. Coates, ed.), p. 263-278. Publ. in Geomorphology, SUNY, Binghamton, N.Y.

El-Ashry, M.T. and Wanless, H.R. (1968). Photo-interpretation of shoreline changes between Capes Hatteras and Fear, North Carolina. Marine Geology 6, 347-379.

Field, M.E. and Duane, D.B. (1976). Post-Pleistocene history of the United States inner continental shelf: Significance to origin of barrier islands. Geol. Soc. Amer. Bull. 76, p. 77-86.

Fisher, J.J. (1962). Geomorphic expression of former inlets along the Outer Banks of North Carolina. M.A. thesis, Univ. N.C., Chapel Hill, 120 p.

Godfrey, P.J. and Godfrey, M.M. (1973). Comparison of ecological and geomorphic interactions between altered and unaltered barrier island systems in North Carolina. In "Coastal Geomorphology" (D.R. Coates, ed.), p. 239-258. Publications in Geomorphology, SUNY, Binghamton, N.Y.

Godfrey, P.J. and Godfrey, M.M. (1974). The role of overwash
    and inlet dynamics in the formation of salt marshes on
    North Carolina barrier islands. Ecology of Halophytes,
    Academic Press, New York, p. 407-427.
Godfrey, P.J. and Godfrey, M.M. (1976). Barrier island ecology
    of Cape Lookout National Seashore and vicinity, North Caro-
    lina, National Park Service Scientific Monograph Series,
    No. 9, 166 p.
Hayes, M.O., Owens, E.H., Hubbard, D.K., and Abele, R.L. (1973).
    The investigation of form and processes in the coastal zone.
    In "Coastal Geomorphology" (D.R. Coates, ed.), p. 11-42.
    Pub. in Geomorphology, SUNY, Binghamton, N.Y.
Hosier, P.E. and Cleary, W.J. (1977). Cyclic geomorphic pat-
    terns of washover on a barrier island in Southeastern North
    Carolina. *Env. Geol. 2.,* p. 23-31.
Jarrett, J.T. (1977). Sediment budget analysis, Wrightsville
    Beach to Kure Beach, North Carolina. *In* Coastal Sediments
    '77, Amer. Soc. Civil Engineers, N.Y., p. 986-1005.
Langfelder, L.J., Stafford, D., and Amein, M. (1968). A recon-
    naissance of coastal erosion in North Carolina. NCSU,
    Raleigh, Dept. Civil Eng. Rept., 127 p.
Langfelder, L.J., French, T., McDonald, R., and Ledbetter, R.
    (1974). A historical review of some of North Carolina's
    coastal inlets. Center for Marine and Coastal Studies,
    NCSU, Raleigh. Rept. No. 74-1, 43 p.
Magnuson, N.C. (1965). Planning and design of a low weir sec-
    tion jetty at Masonboro Inlet, North Carolina. Coastal En-
    gineering Specialty Conference, Santa Barbara, Calif.,
    Chapter 36, p. 807-820.
Moslow, T.F. and Heron, D. (1978). Rates of landward migration
    of Core Banks, North Carolina--Effects on depositional pat-
    terns over past 7000 years. AAPG-SEPM Program, Oklahoma
    City, Oklahoma, p. 98 (abs.)
Nummedal, D., Oertel, G.F., Hubbard, D.K., and Hine, A.C.
    (1977). Tidal inlet variability-Cape Hatteras to Cape Cana-
    veral. *In* Coastal Sediments '77. Amer. Soc. Civil Eng.,
    N.Y., p. 543-562.
Pierce, J.W. (1969). Tidal inlets and washover fans. *J. Geol.*
    61, 230-234.
Pierce, J.W. and Colquhoun, D.J. (1970). Holocene evolution of
    portion of the North Carolina coast. *Geol. Soc. Amer.*
    Bull. 81, p. 3697-3714.
Schwartz, R.K. (1975). Nature and genesis of some storm wash-
    over deposits. U.S. Army Coastal Eng. Research Center,
    Tech. Memo. 61, 69 p.
Shideler, G.L. (1973). Textural trend analysis of coastal bar-
    rier sediments along the middle Atlantic Bight, North Caro-
    lina. *Sedimentary Geology 9,* p. 195-220.

Sussman, D.R. and Heron, D. (in press). Evolution of a barrier
    island - Shackleford Bank, Carteret County, North Carolina,
    Geol. Soc. Amer. Bull.
U.S. Army, Corps of Engineers (1964). Ocracoke Inlet to Beau-
    fort Inlet, North Carolina. Combined Hurricane Survey In-
    terim Report - Ocracoke Inlet to Beaufort Inlet, and Beach
    Erosion Report on Cooperative Study of Ocracoke Inlet to
    Cape Lookout. U.S. Army Engineers District, Wilmington, N.C.
U.S. Army Corps of Engineers (1973). General design Memorandum.
    Phase I. Hurricane-wave protection and beach erosion con-
    trol, Brunswick County, North Carolina. Beach projects,
    Yaupon and Long Beach segments, U.S. Army Eng. District,
    Wilmington, N.C.
Vallianos, L. (1970). Recent history of erosion at Carolina
    Beach, North Carolina. Proc. Twelfth Coastal Eng. Conf.,
    Washington, D.C. Chptr. 77, p. 1223-1242.
Vallianos, L. (1975). A recent history of Masonboro Inlet,
    North Carolina. *In* "Estuarine Research" (L.E. Cronin, ed.).
    Academic Press, N.Y.

BARRIER ISLAND DEVELOPMENT DURING THE HOLOCENE RECESSION, SOUTHEASTERN UNITED STATES

George F. Oertel

Institute of Oceanography
Old Dominion University
Norfolk, Virginia

Barrier islands and marsh lagoons along the mesotidal coast of southern South Carolina, Georgia, and northern Florida have reoccupied the positions of former Pleistocene shores and lagoons. During the late Holocene recession, barriers were welded onto the Silver Bluff (Pleistocene) barriers and the sea reoccupied relict embayments. In areas where rivers contribute sediment to the coast, the Holocene barriers are not welded onto the Silver Bluff (Pleistocene) but are oblique to the relict shoreline. Initial shore orientations and lagoonal drainage patterns were determined by Pleistocene topographic conditions. Adjustments after Holocene welding are partially illustrated by beach ridges and truncations between sets of beach ridges. Sediment eroded from ocean beaches is transported toward inlets, where it caused constriction of inlets and expansion of inlet shoals.

Prevailing longshore currents do not exist in this section of the Coastal Plain Physiographic Province, although seasonal northeast storms apparently produce a residual southerly flow on an annual basis. Reversals in sediment transport by wave refraction are also caused by inlet shoals.

Modification and development of Holocene barriers was apparently controlled by: (1) Holocene rise of sea level, (2) the relict Pleistocene topography, (3) inlet tidal-current patterns, (4) seasonal storms, and (5) availability and supply of sediment.

INTRODUCTION

The Sea Island section of the Coastal Plain Physiographic
Province illustrates a variety of Holocene sequences (Fenneman
1938; Todd 1968). Work by Bruun (1966); Hayes (1975); Oertel
(1974, 1975a); and Nummedal et al. (1977) has demonstrated
that the relative forces of inlet currents versus wave currents
have a significant influence on inlet form and development. In
the southeastern United States a continuum of inlet forms
exists between the wave-dominated margins and the tide-dominated
center of the Georgia embayment (FitzGerald and FitzGerald,
1977; Nummedal et al., 1977). Because the development of tidal inlets has a major influence upon development of barrier
islands (Johnson 1919; Hoyt and Henry, 1967), the regional
change in inlet types may affect the form and development of
barrier islands. This is clearly demonstrated along the southeastern coast of the United States. Barrier islands along the
wave-dominated coasts of the Outer Banks of North Carolina are
long and narrow, and migrate in the direction of the prevailing
longshore current. Barrier islands in the tide-dominated center of the Georgia embayment are relatively short and have been
relatively immobile during their 2,000-4,000 year Holocene history. Although the recession of the Holocene shoreline began
at the edge of the continental shelf approximately 18,000 BP,
only the most recent 2,000-4,000 years are represented by subaerial Holocene deposits (Fig. 1). With the return of the
Holocene sea, the shoreline migrated westward across the continental shelf and eventually came in contact with the relict
Silver Bluff (Pleistocene) shore (Pilkey and Field, 1972; Swift
et al., 1971; Swift et al., 1972).
     This paper attempts to review the developmental patterns in
Georgia barrier islands from the time when the Holocene seas
reoccupied the relict Silver Bluff (Pleistocene) bays and lagoons. Archeological evidence by DePratter and Howard (1977)
suggests that this occurred approximately 4,500 BP.
     Two obvious patterns developed when the Holocene sand sheet
merged with relict Silver Bluff (Pleistocene) deposits. Areas
that were not influenced by major rivers have Holocene beaches
"welded" directly to relict Silver Bluff shore. Areas of Silver Bluff shore (Fig. 1) that are influenced by major rivers
are fronted by 3 to 7 km of marsh fringed with Holocene barrier
islands. These sediment prisms are deltas of the late Holocene
rivers. Both the Savannah and Altamaha Rivers have deltas
formed by barrier beaches oriented oblique to the Silver Bluff
(Pleistocene) shoreline.

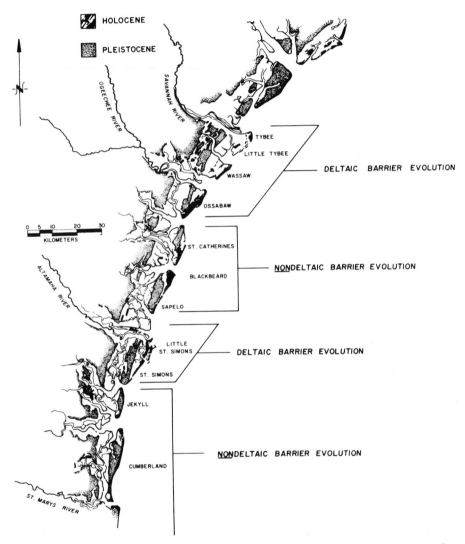

Fig. 1. Subaerial distribution of Holocene deposits and their relationship to the relict Silver Bluff (Pleistocene) shoreline. (Modified after DePratter and Howard, 1977).

OBSERVATIONS

At stable coastal areas, foredunes form at spring tide berms
where wind-blown sand is trapped by mounds of detritus that has
been washed from bays and marshes (Oertel and Larsen, 1976;
Ranwell 1972). Pioneer plants take hold in the foredunes and
spread by rapid rhizomatous development. Dune plants grow upward at a rate greater than burial and thus emerging dunes generally do not migrate but expand in place. Expanding foredunes eventually coalesce and form a ridge over the spring tide
berm. Coastal dune ridges are therefore the record of former
spring tide shorelines, and truncated ridges are the erosional
interruptions to the sequence. The mere absence of dune ridges
is insufficient to suggest a hiatus; erosional indicators must
be used in conjunction with non-ridged areas. The cross-
cutting and juxtaposed relationships of dune ridges provide a
valuable tool for interpreting the Holocene evolution of barrier islands.

In areas not influenced by major rivers, barrier islands of
the Atlantic coast of southern South Carolina, Georgia, and
northern Florida have Pleistocene cores and Holocene beaches.
The Holocene beaches are generally thin in the center-island
areas, and broader adjacent to the inlets (Fig. 1). Holocene inlets are considerably more constricted today than they were
2,000-4,000 years BP (Fig. 2). Dune ridges in the center-
island areas generally parallel the modern shoreline. However,
dune ridges adjacent to the inlets illustrate sets of recurved
patterns that are often separated by truncations. The sequen-
tial development of these sets illustrates a channelward evolution through time (Fig. 2).

A simple explanation of recurved spit development is applicable to the south end of these barrier islands (Hoyt and
Henry, 1967). However, Holocene development on the north ends
of barrier islands is more complicated and apparently related
to the cyclic development of ebb-tidal deltas (Oertel 1977).
While processes of infilling at the northern and southern margins of tidal channels are different, the net effect was to
produce inlet constriction during the Late Holocene (2,000-
4,000 years BP).

In areas of major river influence, barrier islands are separated from the Silver Bluff (Pleistocene) by a 3-7 km wide
delta composed of hammocks, marshes and tidal channels. Deltas
are located south of each river and attain triangularity in
planview (Fig. 3). The coarse-grained deposits in Holocene
deltas are primarily located at the outer edges of the sediment
prisms forming "marsh-encircled" islands or hammocks and barrier beaches. The innermost portions of deltas are conspicu-
ously devoid of coarse-grained deposits and are composed of

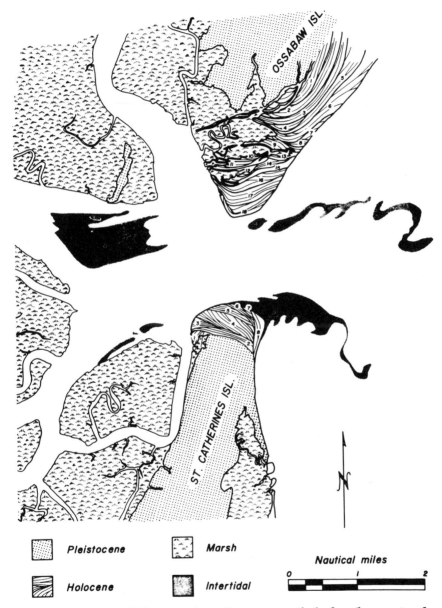

Fig. 2. Map illustrating the sequential development of
Holocene beach ridges, illustrating constriction at the inlet
throat (St. Catherines Sound, Georgia).(Adopted from Oertel
1975b).

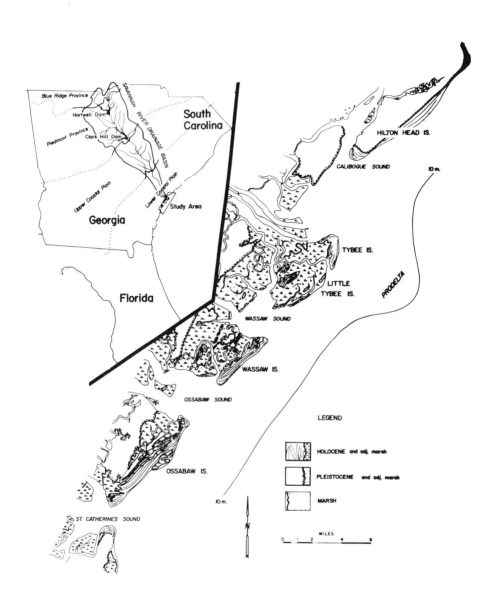

Fig. 3.   Map depicting the Holocene Savannah River delta
with isolated hammocks, marsh and barrier islands.

marsh deposits.   While DePratter and Howard (1977) suggest that
"Hammocks are erosional remnants of previously more extensive
and continuous barrier island beach ridges and spits which have
subsequently been segmented and eroded by meandering tidal
creeks and rivers," the author sees no evidence for the former
existence of an extensive barrier island chain within the ma-
trix of the delta.   Hammocks in the delta are discrete and complete units that are related to overwash processes or accretion
in echelon spits and marshes in conjunction with shifting inlets (Figs. 2 and 4) (Oertel 1975b, 1977; Hubbard 1977; Pickering, et al., in preparation).

EVOLUTION OF TIDE-DOMINATED BARRIER ISLANDS

The history and chronology of the tide-dominated coastal
area of the southeastern United States during the last 4,500
years was initially controlled by the physiography of the existing Pleistocene shore that formed 25,000 to 36,000 years BP
(Hails and Hoyt, 1968; Hoyt et al., 1968).   At the end of the
Wisconsin glaciation, approximately i8,000 years BP, sea level
began rising and the shoreline began receding from the edge of
the continental shelf (-100 m depth, Fairbridge 1961).   At
approximately 5,000 BP the sea level was 2 to 3 meters below
its present level (Shepard and Curray, 1967; Milliman and
Emery, 1968), and the Holocene shoreline came into contact with
steeper slopes created by the relict Silver Bluff· shoreface.
Low-lying areas behind the relict Silver Bluff shore were reflooded by the rising Holocene sea to form lagoons (Hoyt 1967).
These were not "open-water lagoons," as is common behind the
Outer Banks of North Carolina, but "complex lagoons" composed
of marshes, tidal channels, and sounds.
The exchange of water between the newly reoccupied lagoons
and the Atlantic Ocean established the present tide-dominated
area over the inner shoreface of the central Georgia embayment.
Except during storms, inlet tidal currents became the predominant force dispersing sediments over the inner 5-10 km of
shoreface (Fig. 5).   Waves and wind-driven currents controlled
the dispersion of sediments seaward of this zone and at the
foreshore.

Nondeltaic Barrier Islands

In areas not influenced by rivers, the major sediment supply for the retreating shore was the shore itself.   Without a
large sediment supply, shoreface connected shoals were isolated
on the shelf (Duane et al. 1972; Swift 1975).   Swift (1975)

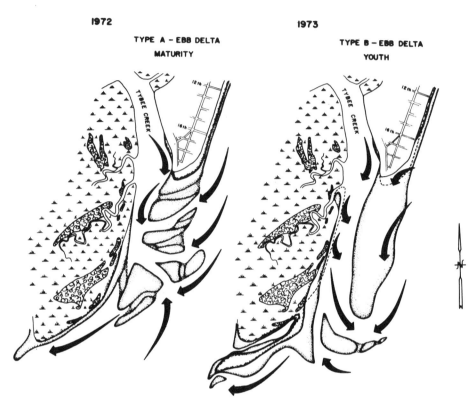

Fig. 4. Map of the Tybee Creek inlet system illustrating enechelon spit (hammock) and marsh development. Several mechanisms of enechelon spit and marsh development are active along the southeast Georgia embayment (Adopted from Oertel 1977).

termed these features shoal-retreat massifs. Shoreline retreat was accomplished primarily by overwash processes (Godfrey and Godfrey, 1973). When the retreating Holocene shoreline came into contact with the higher elevations of the Silver Bluff (Pleistocene), overwash processes were inhibited and sediment transport was controlled by tidal and wave-induced longshore currents. In low-lying areas the Holocene shore continued to retreat by overwash processes, but adjacent to Pleistocene scarps, eroded material was transported toward the ends of islands. Deposition at the southern tips of islands occurred as echelon spits and marshes that curved into inlets (Figs. 2 and 4). Dune ridges accent the surfaces of these spits. Depositional sets on the north ends of islands were generally not

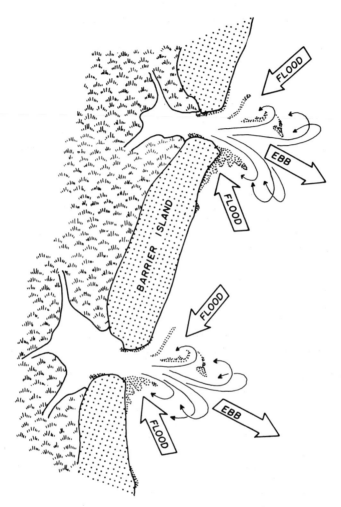

Fig. 5. Map showing shoreface circulation patterns along tide-dominated coasts. (Adopted from Oertel 1975b).

separated by marshes. Dune ridges illustrating the evolution of depositional sets were parallel; but adjacent sets were oriented oblique to each other, suggesting a new direction of shoreline development. Oertel (1977) noted a relationship between the growth at the ends of modern barrier islands and the current deflection produced by the marginal shoals of ebb deltas. When the shoals were attached to the foreshore of the island, flood currents were deflected offshore and the inlet margin of the island eroded rapidly (Fig. 6). When shoals were separated from the island, accretion occurred adjacent to the end of the island. Marginal shoals apparently undergo a cyclic development, producing repeated attachments and separations from the foreshore. This cyclic pattern was active throughout the late Holocene and produced alternating sets of beach ridge development on the northern ends of tide-dominated barrier islands (Oertel 1975b).

As water ebbed from the newly reoccupied Silver Bluff inlets, large ebb deltas formed that interrupted the flow of coastal currents. The inlet tidal currents affected and were affected by the ebb deltas, as they are today. While the duration of the ocean tides is symmetrical, the exchange of water between the ocean and "complex lagoons" is asymmetrical. The processes causing the asymmetrical exchange from modern "complex lagoons" has been described by Hubbard (1977), and Nummedal and Humphries (in press). The effect of this asymmetry is to produce ebb-dominated zones along the main axial channels of ebb deltas and flood-dominated zones between the beaches and the margins of ebb deltas (Figs. 5 and 7). The flood-dominated zones are particularly important for transporting sediment from the shoreface adjacent to the center of islands to the foreshore areas at the ends of islands. As pointed out above, the Holocene evolution of the tide-dominated barrier islands illustrates shore-normal channels and major accretionary sections at the ends of each island. Inlets and islands do not illustrate an offset in a downdrift direction. This indicates that late Holocene circulation patterns were similar to modern circulation patterns at inlet-tide dominated coasts.

Diverging patterns of wave-induced longshore flow and flood tidal currents caused the retreating Holocene sediment prisms to develop more rapidly adjacent to the inlets than in the central island areas (Oertel 1975b). The constriction of inlet throats during the past 4,000 years was not an equilibrium response to variations in lagoon size (O'Brien 1931, 1969), but rather the effect that the inlet-tide dominated shoreface had on the merger between the high elevations of a relict shore and the retreating Holocene sediment prism (Fig. 7).

Deltaic Barrier Islands

In areas influenced by river outflow, rivers have modified the system in two ways. Prodeltas associated with relict Pleistocene rivers produced lobate topographic highs seaward of the Silver Bluff shoreline. As the Holocene sea returned, the rivers also supplied the beaches with an additional source of sediment to the shoreface system. While the residence time of sand on the shore was not changed, there was an increase in the

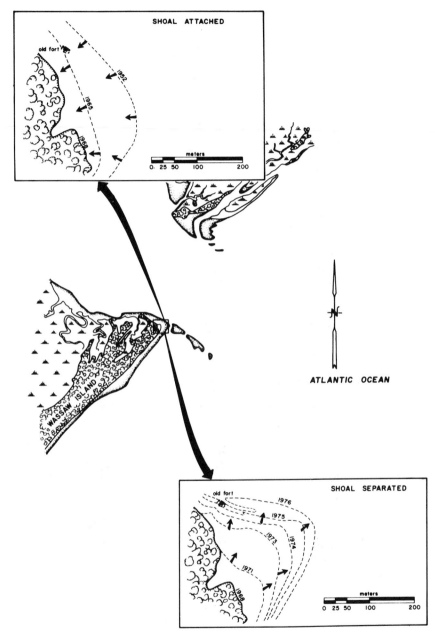

Fig. 6. Map of the north end of Wassaw Island, Georgia illustrating erosional and accretional modes of inlet-island development, during different stages of ebb delta evolution. (Adopted from Oertel, 1977).

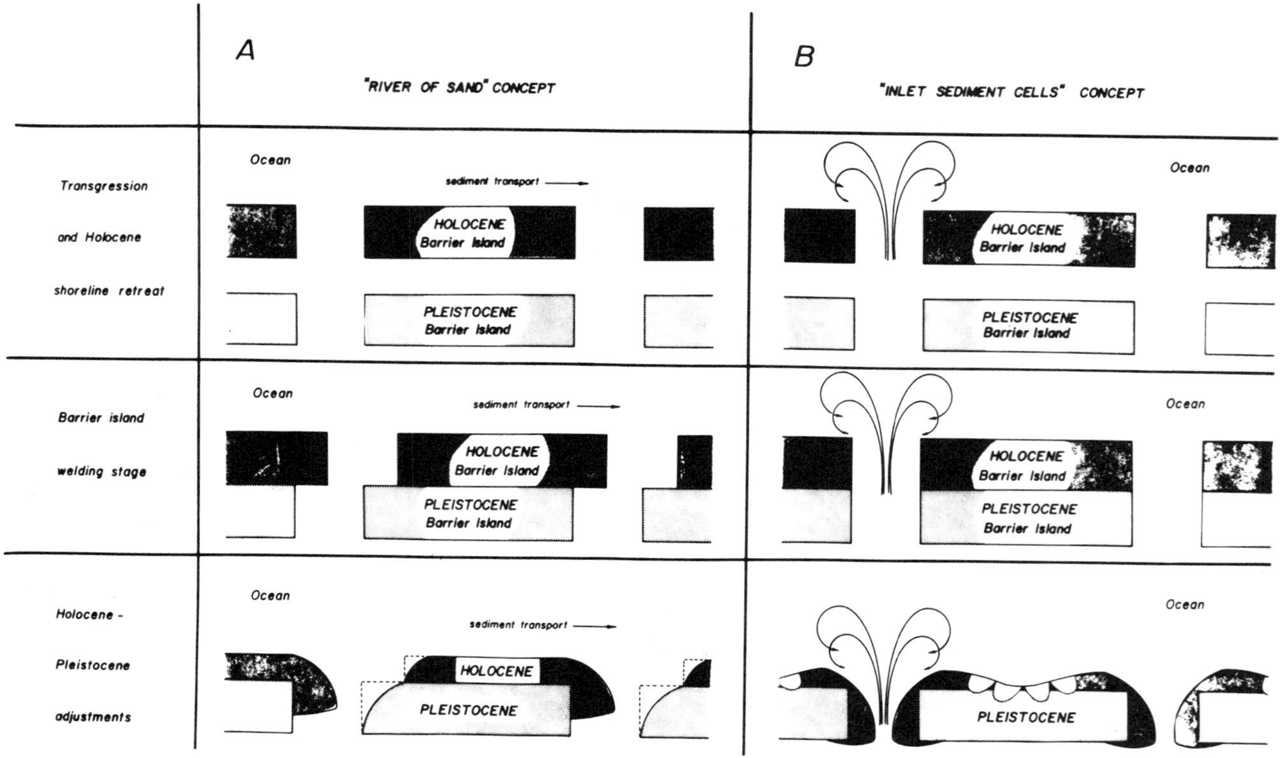

Fig. 7. Hypothetical models of barrier retreat during later Holocene transgression. The diagrams delineate two ideal systems not influenced by rivers. Islands and inlets are offset in a down-current direction in the "River of Sand" concept.

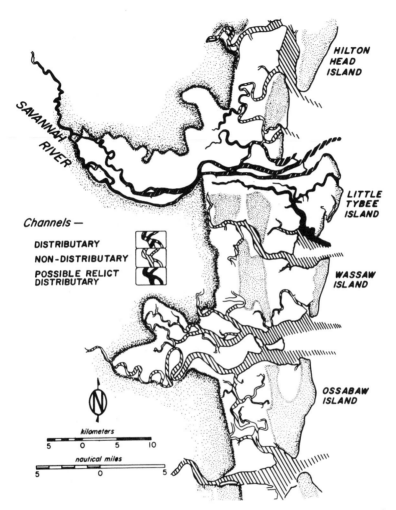

Fig. 8. Map illustrating relict distributaries of the Savannah River and non-distributary channels that have dissected the retreating Holocene shoreline.

quantity of sediment needed to be moved in order to produce shoreline recession. Thus, the rates of shore retreat adjacent to fluvial systems is less than that observed at beaches not affected by major rivers. With the additional supply of sediment, shoreface-connected shoals were less likely to be isolated on the shelf and were more likely to accrete at flow convergences.

Holocene barrier development adjacent to major river systems began approximately 5,000 years BP when sea level was

2-4 meters lower (Shepard and Curray, 1967; Milliman and Emery, 1968; DePratter and Howard, 1978), and the retreating Holocene sediment prism collided with the prodeltas and shoals of the relict river systems. Thin barrier beaches perched on these relict highs were supplied with sediment from river distributaries and retreating Holocene shorelines. Small barrier beaches formed at the confluences of the sea and the numerous small distributary channels (Fig. 8). The additional sediment supply from the fluvial source encouraged a still-stand of the retreating shore 3-7 km seaward of the relict Silver Bluff (Pleistocene) shore.

Initially, a semicontinuous chain of barrier islands did not form due to the relatively low longshore transport of sediment, and the low supply of sediment from "drowning" rivers. In the Savannah River area, ceramic archaeological findings (DePratter and Howard, 1977) suggest that this early developmental period occurred from 3,700 to 1,000 years BP.

Between 675 and 1,000 years BP the equilibrated shoreface and Savannah River became more active suppliers of sediment to the system, and barrier spits formed adjacent to the confluence of the main re-entrant channel and the sea. Stratified circulation at the river entrance allowed ebb deltas to form close to inlet throats and permitted a more rapid transfer of sediment to adjacent beaches. This area of sediment accumulation developed bidirectionally, similar to the patterns for non-river influenced barriers described above. The northward growth was sporadic and apparently related to inlet tidal currents and cyclic shifts in ebb-delta shape (Oertel 1977), whereas the southward growth was relatively continuous and apparently influenced to a great extent by the seasonally predominant flow of wave-induced foreshore currents. While deltaic barrier development was progressing, the relict delta surface behind the barriers was being reflooded and modern marsh drainage patterns began evolving. During the recent 78-year history (1897-1975) of the Georgia coast, the deltaic barriers of the Savannah and Altamaha River systems were still accumulating sediment (Oertel and Chamberlain, 1975; Oertel 1978), whereas non-deltaic barriers experienced net erosion.

CONCLUSIONS

Approximately 4,000 years BP the retreating Holocene shoreline penecontemporaneously collided with relict shoals and prodeltas of Pleistocene rivers and the relict Silver Bluff foreshore.

In areas not influenced by major rivers the reflooding of Pleistocene lagoons established tide-dominated inlet circulation over the shallow shoreface. Flow over the shoreface diverged away from the center island areas and toward the ends of islands. Holocene depositional patterns reflect these circulation patterns.

The circulation in river-dominated areas was controlled initially by the confluences of stratified distributaries and coastal currents. Small barrier beaches formed at the confluences of these distributaries and the sea. In areas where the retreating Holocene shoreline collided with topographic highs of relict river deltas, Holocene barrier islands developed seaward of the Silver Bluff contact. The added supply of sediment from equilibrated modern rivers enhanced deltaic barrier development, some 3 to 7 km seaward of the Silver Bluff foreshore.

ACKNOWLEDGMENTS

The author profited from discussions with V.J. Henry and appreciated his review of the manuscript. R. Brokaw assisted in figure preparation. The author appreciated the suggestions of C.F. Chamberlain.

REFERENCES

Bruun, P. (1966). Tidal inlets and littoral drift: Stability of coastal inlets, v. 2, 193 p.

DePratter, C.B. and Howard, J.D. (1977). History of shoreline changes determined by Archaeological dating: Georgia coast, U.S. *Transactions Gulf Coast Assoc. of Geol. Societies* XXVII, p. 252-258.

DePratter, C.B. and Howard, J.D. (1978). Archaeological evidence for Post-Pleistocene low stand of sea level on southeastern United States coast. (Abs.) Program A.A.P.G.-S.E.P.M. annual convention, Oklahoma City, Oklahoma.

Duane, D.B., Field, M.E., Meisburger, E.P., Swift, D.J.P., and Williams, S.J. (1972). Linear shoals on the Atlantic inner continental shelf, Florida to Long Island. *In* "Shelf Sediment transport: process and pattern"(D.J.P. Swift, D.B. Duane, and O.H. Pilkey, eds.), p. 447-498. Dowden, Hutchinson & Ross, Stroudsburg, Pennsylvania.

Fairbridge, R.W. (1961). Eustatic changes in sea level. *In* "Physics and Chemistry of the Earth," p. 99-185. Pergamon Press, N.Y.

Fenneman, N.M. (1938). Physiography of eastern United States, McGraw-Hill, New York and London, 44 p.

FitzGerald, D.M. and FitzGerald, S.A. (1977). Factors influencing tidal inlet throat geometry. *In* "Coastal Sediments '77," Fifth Symposium of the Waterway, Port, Coastal and Ocean Division of the American Society of Civil Engineers, Charleston, South Carolina, p. 563-581.

Godfrey, P.J. and Godfrey, M.M. (1973). Comparison of ecological and geomorphic interactions between altered and unaltered barrier island systems in North Carolina. *In* "Coastal Geomorphology" (D.R. Coates, ed.), p. 239-257. Univ. of N.Y.

Hails, J.R. and Hoyt, J.H. (1968). Barrier development on submerged coasts: problems of sea-level changes from a study of the Atlantic Coastal Plain of Georgia, United States and parts of the east Australian Coast. Zeitschrift für Geomorphologie, Annals of Geomorphology, new series, Supplement v. 7, p. 24-55.

Hayes, M.O. (1975). Morphology of sand accumulation in estuaries: an introduction to the Symposium. *In* "Estuarine Research," v. 2, p. 3-22. Academic Press, N.Y.

Hoyt, J.H. (1967). Barrier island formation. *Geol. Soc. America Bull. 78*, p. 1125-1136.

Hoyt, J.H. and V.J. Henry (1967). Influence of island migration on barrier-island sedimentation. *Geol. Soc. America Bull. 78*, p. 77-86.

Hoyt, J.H., Henry,Jr., V.J., and Weimer, R.J. (1968). Age of late-Pleistocene shoreline deposits, coastal Georgia. *In* "Means of correlation of Quaternary successions." Cong. of Int. Assoc. for Quaternary Research, v. 8, p. 381-393.

Hubbard, D.K. (1977). Variations in tidal inlet processes and morphology in the Georgia embayment. Tech. Rpt. No. 14-CRD, Coastal Research Div., Univ. of South Carolina, Columbia.

Johnson, D.W. (1919). Shore processes and shoreline development. Hafner Publishing Co., 584 p.

Milliman, J.D. and Emery, K.O. (1968). Sea levels during the past 35,000 years. *Science 162*, 1121-1123.

Nash, G.J. (1978). Historical changes in the mean high water shoreline and nearshore bathymetry of south Georgia and north Florida. Unpubl. M.S. thesis, 149 p., Univ. of Georgia, Athens.

Nummedal, D.and Humphries, S.M. (1977). Hydraulics and Dynamics of North Inlet, South Carolina, 1975-76. U.S. Army Corps of Engineers, Coastal Engineering Research Center (in press).

Nummedal, D., Oertel, G.F., Hubbard, D.K., and Hine, A.C. (1977), Tidal inlet variability Cape Hatteras to Cape Canaveral. *In* "Coastal Sediments '77," Fifth Symposium of Waterway, Port, Coastal and Ocean Division of the American Society of Civil Engineers, Charleston, S.C., p. 543-562.

O'Brien, M.P. (1969). Equilibrium flow areas of inlets on sandy coasts. *J. of Waterways and Harbors Div.*, Proc. of Amer. Soc. of Civil Eng. WWI *95*, p. 43-52.

Oertel, G.F. (2974). Residual currents and sediment exchange between estuary margins and the inner shelf, Southeast coast of the United States. *In* Symposium volume on inter-relationships of estuarine and continental shelf sedimentation. Bordeaux, France: Memoires de l'Institut de Geologie du Bassin d'Aquitaine, p. 135-143.

Oertel, G.F. (1975a). Ebb tidal deltas of Georgia estuaries. *In* "Estuarine Research," v. 2, Geology and Engineering, (L.E. Cronin, ed.). Academic Press, N.Y., 587 p.

Oertel, G.F. (1975b). Post-Pleistocene island and inlet adjustment along the Georgia coast. *J. Sed. Petrology 45*, p. 150-159.

Oertel, G.F. (1977). Geomorphic cycles in ebb deltas and related patterns of shore erosion and accretion. *J. Sed. Petrology 47*, 1121-1131.

Oertel, G.F. (1978). Report on the historical sediment budgets of the Savannah and Altamaha Rivers, Georgia (unpublished). Georgia Dept. of Natural Resources, Atlanta. 30 p.

Oertel, G.F. and Chamberlain, C.F. (1975). Differential rates of shoreline advance and retreat as coastal barriers of Chatham and Liberty counties, Georgia. *Trans. Gulf Coast Assoc. of Geol. Soc. 25*, p. 383-390.

Oertel, G.F. and Larsen, M. (1976). Developmental sequences in Georgia coastal dunes and distributions of dune plants. *Bull. Georgia Acad. Science 34*, p. 35-48.

Pickering, S.M., Henry, V.J. and Giles, R.T. (in prep.). Williamson Island: a new island on Georgia's coast.

Pilkey, O.H. and Field, M.E. (1972). Onshore transportation of continental shelf sediment: Atlantic southeastern United States. *In* "Shelf sediment transport: process and pattern," (D.J.P. Swift, D.B. Duane, and O.H. Pilkey, eds.), p. 429-446. Dowden, Hutchinson & Ross, Stroudsburg, Pennsylvania.

Ranwell, D.W. (1972). Ecology of salt marshes and sand dunes. Chapman and Hall, Ltd., London, 258 p.

Shepard, F.P. and Curray, J.R. (1967). Carbon-14 determination of sea level changes in stable areas. *In* "The Quaternary history of the ocean basins," p. 283-291. Cong. of Int. Assoc. for Quaternary Research, Boulder, Colorado.

Swift, D.J.P. (1975). Tidal ridges and shoal retreat massifs. *Marine Geology 18*, 105-134.

Swift, D.J.P., Kofoed, J.W., Saulsbury, F.P., and Sears, P. (1972). Holocene evolution of the shelf central and south Atlantic shelf of North America. *In* "Shelf sediment transport: process and pattern (D.J.P. Swift, D.B. Duane, and O.H. Pilkey, eds.), p. 499-574. Dowden, Hutchinson & Ross, Stroudsburg, Pennsylvania.

Swift, D.J.P., Sanford, R.B., Dill, Jr., C.E., and Avignone, N.F. (1971). Textural differentation on the shoreface during erosional retreat of an unconsolidated coast, Cape Henry to Cape Hatteras, western North Atlantic shelf. *Sedimentology 16*, 221-250.

Todd, T.W. (1968). Dynamic diversion--influence of longshore current-tidal flow interaction on chenier and barrier island plains. *J. Sed. Petrology 38*, p. 734-746.

BARRIER ISLAND EVOLUTION AND HISTORY OF MIGRATION, NORTH CENTRAL GULF COAST

Ervin G. Otvos, Jr.

Gulf Coast Research Laboratory
Ocean Springs, Mississippi

Preliminary results of core-drilling on the Mississippi-Alabama barrier islands, with several lines across the Mississippi Sound, indicate that the islands evolved by shoal-bar aggradation, probably not earlier than three to four thousand years ago. Alternate theories of barrier island formation (spit segmentation or mainland dune-ridge engulfment) did not seem valid in the subject area. With the exception of eastern Dauphin Island, open marine nearshore deposits underlie the present islands. Toward the seaward areas of the Sound, brackish bottom sediments are underlain by higher salinity deposits. Full-salinity conditions are not always expected even in an originally unbarred coastal setting, regardless of its later conversion to lagoonal conditions. At the time of the Holocene transgression, eastern Dauphin Island represented a higher, Pleistocene ground that was only veneered by Holocene beach and dune deposits. By "capturing" the sand that arrives from the Alabama mainland shore through current and drift processes via the Mobile Bay ebb-tidal delta, and steering it westward along its south shore, eastern Dauphin Island probably played an important role in originally determining the general offshore position of the whole barrier island chain.

Historical charts since the 18th century record the (re)emergence of numerous small and a few larger islands from shoals. Their occurrence was most apparent in the southern Chandeleur and Mobile Bay ebb-tidal delta areas after storm destruction of pre-existing islands. Landward migration rates of the Chandeleur Islands ranged between 0.6-2.1 km/century. These islands are shifted by storm overwash and sediment transfer through inlets. This migrational process is also accomplished by the post-storm island reemergence in positions landward of those occupied prior to a hurricane. For the Mississippi-Alabama

*barrier islands, spit growth and reattachment plays a major role
in island progradation. Westward accretion values for four of
the islands ranged between 1.3-7.4 km between 1848-1974. Hurricane destruction and segmentation has been episodic but has
played an essential role in the evolution of all the islands.
Historic data indicate that the final separation of Dauphin
and Petit Bois Islands occurred between 1740-1766; possibly
during the 1740 hurricane.*

INTRODUCTION

   This paper summarizes preliminary results, obtained over a
five year period from some seventy coreholes in the Mississippi
Sound and adjacent island and mainland areas. The data has been
amassed to study the Holocene evolution of the Mississippi-
Alabama coastal and nearshore zone. At the same time, evaluation of historic data from the early 18th century to the present allowed the correction and updating of earlier information
(Otvos 1970a,b) on the migration patterns of these areas and
the Chandeleur Islands. It was also used to document the (re)-
emergence of some islands from submerged bar or shoal positions.
This information has a significant bearing on theories of barrier island genesis and subsequent evolution.

STUDY METHODS

Drilling and Sediment Analysis

   Rotary drilling on land, in swamps by swamp-buggy and
over water by barge allowed almost continuous sampling of uncontaminated cores from the Holocene section and underlying
Late Pleistocene units. Samples were not obtained by the less
reliable auger and wash-boring methods. Grain size analysis was
performed by the standard hydrometer and Ro-Tap sieving methods.
Folk statistical values were used in calculating sorting and
other parameters. Foraminifers, which are sensitive to facies
conditions, were by far the most useful fossil group in determining depositional facies. In contrast to molluscans and various microfossils (ostracods, diatoms, etc.), foraminifers
were abundant in almost every sample. Four biofacies groups,
gradational between each other, were identified in the Holocene
brackish and marine deposits. Very few of the radiocarbon dates
were from uncontaminated and non-reworked sample material.

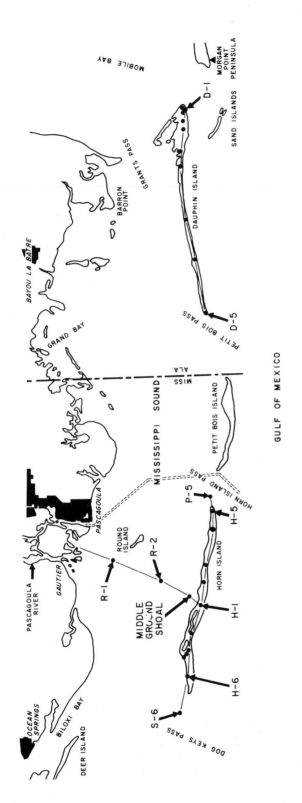

Fig. 1. Mississippi-Alabama barrier islands, eastern area. Black dots: coreholes.

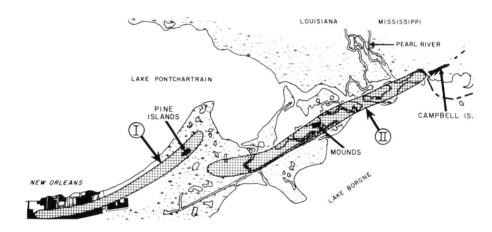

Fig. 2. *Mississippi-Alabama barrier islands, western area.*
Black dots: coreholes. I: New Orleans Holocene barrier trend;
II: Hancock-Sauvage Holocene barrier trend.

Analysis of Historical charts

If the limitations of historical charts are well understood,
such studies provide useful information about former hydrographic and topographic conditions that existed prior to the use
of detailed surveys. Several dozen original and copied 18th
century charts were studied in the Historic New Orleans Collection, the Mississippi State Archives, and the Mobile Public
Library.

While the best 18th century maps were fairly rudimentary,
the general position, size, outlines and number of the islands
were indicated and yielded useful information. Problems arose
when charts were based on earlier ones so that the information
provided sometimes showed conditions that existed 20-50 years
prior to their publication. The first reliable details of the
Mississippi-Alabama nearshore area are shown on the 1847-56
U.S. Coast Survey charts. Unfortunately, the next surveys were
not made until 1916-17. Several intervening published charts
reflected conditions that were outdated, sometimes by half a
century. Following the 1917 charts, published maps still lagged
behind actual field conditions by 10-25 years. Minor changes

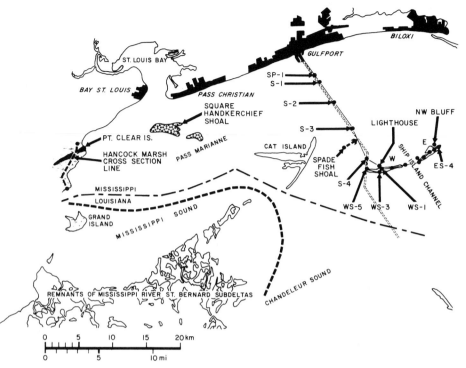

in the configuration of shoals and islands and the correct
identification of these features are still occasionally missing
from the regularly updated, published USCGS charts, as indicated by comparisons with aerial photographs.

In assessing the dimensions of maximum shore changes that
occurred on the Mississippi-Alabama barrier islands during the
past 130 years, it has been necessary to rely heavily on chart-
comparisons for that period, provided by Waller and Malbrough
(1976), and Hardin et al. (1976).

GEOGRAPHIC AND HYDROLOGIC SETTING

The six coast-parallel barrier islands, 3.3-17.2 km off the
Mississippi-Alabama shore (Figs.1 and 2), range in length between 3.5 km (East Ship) and 24 km (Dauphin); and in width between a few hundred meters to 1.5 km. Beaches, instead of
marshes, formed along their Sound shores, due to the width of
the Mississippi Sound and depth of nearshore Sound waters.
Ponds with marshes do occur in the island interiors. The barrier dunes are 0.5-3 m high, with a maximum of 4-6 m. Precipi-
tation dunes in southeastern Dauphin Island reach 14 m. Sand
shoals flank the islands at 0.3-2.7 m depth, representing sites
of earlier island segments. Shore-parallel and sub-parallel,
sometimes arcuate, bars skirt the southern island shores, while
often two or three intersecting sets of sandwaves-megaripples
dominate shoals off the north shores.

Maximum water depths in the 112 km long Mississippi Sound
are close to 6 m; in the passes 6-10 m; and in Mobile Bay entrance channel 15-19 m. Due to fairly abundant fresh water
runoff, largely from the six major streams, salinities for major portions of the Sound range from 2-20 ppt during most of
the year. Only in the passes and certain central areas do salinities rise above 28 ppt, usually during the dry season
(Eleuterius 1977). The south-southeastern wind component is
important in the region almost throughout the whole year, especially in the summer and spring (Eleuterius 1974). This accounts for the dominant westward-moving littoral drift, estimated as 65-196 thousand yd^3 per year   (U.S.   Army,   1971,
p. e7-e8).

Strong tidal currents (occasionally exceeding 0.9 m/sec;
U.S. Army Corps of Engineers, Mobile District, written comm.)
tend to excavate the Mobile Bay entrance channel and limit the
steady westward progradation of Mobile Point Peninsula. Tidal
and wind-driven currents with breaker and swash action built
the broad ebb-tidal delta south of the Pass, which provides the
route for sand movement to Dauphin Island. Smaller deltas between the other islands perform an identical function (Otvos
1975).

The Chandeleur island-shoal chain forms a 77 km, north-south
directed gentle arch between the Gulf and the brackish, 27-36
km wide, 2-8 m deep Chandeleur Sound. In the north, the islands
are more continuous, larger, wider (maximum 1.6 km) and higher
(maximum 5.5 m dune heights) than in the south, where shoals
and small islands predominate. Mangrove swamps and marshes have
developed along the Sound shores.

The great (1-8 km) widths of shoals on the Soundward side
of the islands accounts for the low energy conditions. These
fragile islands continuously change their dimensions, shapes,
and number. Due to the orientation of the island chain, the
north-northwest winds and waves that dominate between September
and February also play a role in transporting foreshore and
nearshore sediments. Despite the absence of major tidal passes,
underwater sand lobes are prominent off both the sound and the
seaward inlet ends, especially in the southern Chandeleur area.
Mainland sediments cannot reach the islands; their only sediment source is from littoral drift along the islands and wave-
excavated Gulf-bottom sands.

GEOLOGICAL EVOLUTION

Late Pleistocene

The pre-Holocene land surface in the subject area was underlain by alluvial-fluvial (Prairie Formation), estuarine-to-open
nearshore marine (Biloxi Formation) and coastal barrier ridge
(Gulfport Formation) deposits (Otvos 1972, 1975, 1976b). At
three locations on the future sites of Deer, Round, and eastern
Dauphin Islands, the land surface rose slightly above the surrounding area. These locations are underlain by semi-consolidated,
limonitic, and humate-impregnated sands and silty sands (Fig. 7),
interpreted as links in the chain of the coastal barrier ridge.
This would form an alignment with the Gasque-Gulf Shores segment, behind the southeast Alabama mainland shores. In Holocene
times the Mobile Point (Morgan) Peninsula barrier spit attached
itself to the seaward side of that segment. Finite radiocarbon
dates from woody Pleistocene samples in eastern Dauphin Island
(Fig. 4) were caused by ground water and other contamination. A
more recent (7,7000 yrs BP) date from humate matter in corehole
No. 4, resulted from even more recent contamination (Otvos 1976a).
Additional finite Pleistocene dates from Ship and Horn
Islands (Fig. 8) are also suspect of contamination, in view of
a >40,000 yrs BP date from peat at 13.1 m below sea level,
taken north of East Ship Island. In the cores the Pleistocene
sediments are mostly light gray, with occasional oxidized pale-
to-dark yellow, brown and orange streaks and spots. Silty-muddy
deposits in the Pleistocene are usually much better consolidated
than the overlying Holocene sediments.
At certain locations, Late Pleistocene stream entrenchment
caused considerable relief of the present Mississippi-Alabama
mainland shore. A limited acoustic survey in the Horn-Petit Bois
area by the U.S. Corps of Engineers (Otvos 1976c), however,
does not indicate that the present passes were inherited from
such stream channels. Narrow valleys were apparently located
under the present central island segments, instead of beneath
the passes. An early Pascagoula River channel under central
Horn Island (probably between coreholes #1 and #2; Fig. 8) at
the same location has also been recorded by another acoustic
reflection survey (Curray and Moore 1963, Fig. 7, p. 138).

Holocene Transgressive Phase

Woody material, immediately underlying the first Holocene
brackish deposits in coreholes H-2 and R-1 (Fig. 6) indicate
that sea level reached the -11.7 m elevation after 7300 yrs BP.
In the Biloxi Back Bay (Otvos 1976b, p. 106) fresh water swamps

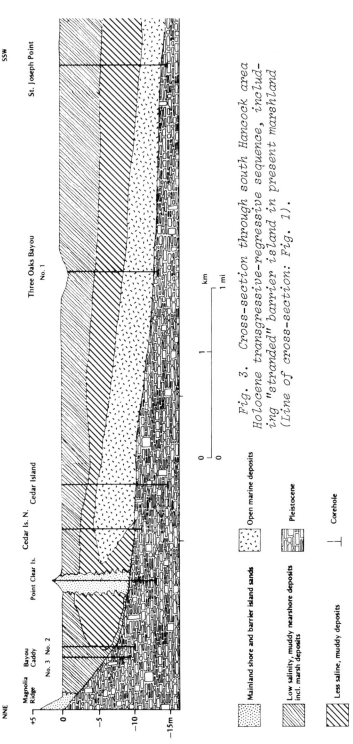

Fig. 3. Cross-section through south Hancock area Holocene transgressive-regressive sequence, including "stranded" barrier island in present marshland (Line of cross-section: Fig. 1).

NNE

Magnolia Ridge

Bayou Caddy
No. 3 No. 2

Point Clear Is.

Cedar Is. N. Cedar Island

Three Oaks Bayou
No. 1

St. Joseph Point

SSW

+5

0

−5

−10

−15m

km

mi

0

1

0

1

Mainland shore and barrier island sands

Low salinity, muddy nearshore deposits incl. marsh deposits

Less saline, muddy deposits

Open marine deposits

Pleistocene

Corehole

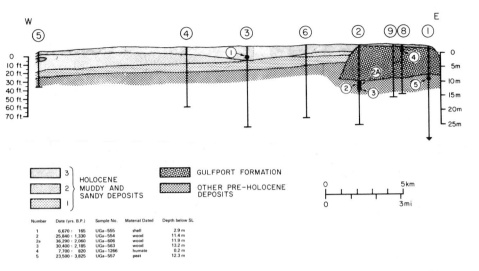

Fig. 4. Holocene units of Dauphin Island: (1) very well to moderately well-sorted sands; (2) moderately to poorly sorted sands; (3) muddy-sandy, clayey deposits. Circles: radiocarbon sample locations.

formed at about 6580 BP at -8.4 to -8.7 m elevations and are overlain at -7 to -7.5 m by brackish sediments, formed about 5700 years ago. Almost all of the present Mississippi Sound area was under water by about five thousand years ago.

Where preserved, brackish, muddy sediments with agglutinated foraminifers represent the early transgression stage, covered by more saline deposits. Open marine foraminifers characterize the muddy, sandy-muddy deposits that underlie the Mississippi-Alabama barrier islands and the adjoining shoal belt. Deposits with such a fauna represent the climax of the mid-late Holocene transgression. Under the present south Hancock marshland, outside the Late Holocene island-shoal belt, open marine sediments were deposited slightly more than 1 km from the mainland (Fig. 3). In the absence of the St. Bernard subdeltas of the Mississippi River and of oligo-mesohaline Lakes Pontchartrain and Borgne (Otvos 1976b, 1978), open Gulf waters reached these shores unhindered. The Pearl River delta front was then located further north, and freshwater influence from that source was less pronounced. In the central parts of the Sound, further to the east, due to a greater fresh water influx, open marine sediments reached only within 7-13 km of the mainland shore (Figs. 5 and 6). Very brackish deposits at the time existed only in bays and within a quite narrow belt along the mainland shore.

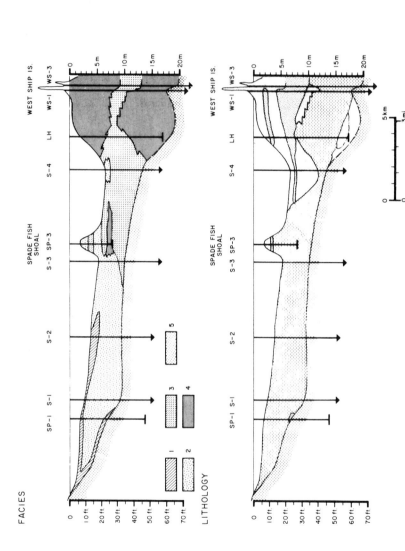

Fig. 5. Lithology and depositional facies of Holocene units between mainland and West Ship Island (Line of cross-section: Fig. 1, Lithology symbols: Fig. 4). Facies: 1 – most brackish; 2 – moderately brackish; 3 – intermediate; 4 – marine; 5 – Pleistocene units

FACIES

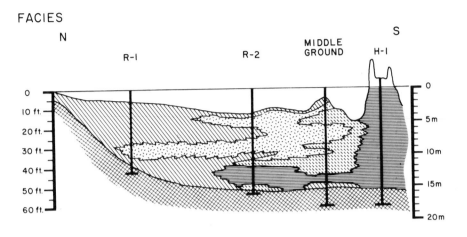

LITHOLOGY

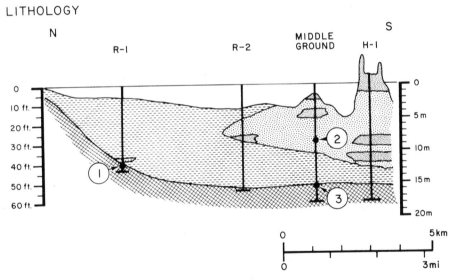

Number	Date (yrs.BP)	Sample No.	Material Dated	Depth below SL
1	7,315 ± 85	UGa-1756	wood	11.7–12.0 m
2	6,765 ± 270	UGa-1757	dispersed plant	8.4– 8.5 m
3	7,825 ± 160	UGa-1758	wood	15.3–15.4 m

Fig. 6. Lithology and depositional facies of Holocene units between mainland and Horn Island (Line of cross-section: Fig. 2; Symbols: Figs. 4 and 5).

Fig. 7. Eroding, limonite and humate-impregnated, loosely consolidated Pleistocene sands, south beach of eastern Dauphin Island (by Fishing Pier).

Barrier Island Formation

The high ground of eastern Dauphin Island probably had a decisive effect in capturing and shunting westward the sand that crossed the developing Mobile Bay ebb-tidal delta platform from Mobile Point (Morgan) Peninsula, which steadily prograded westward. The island became thinly veneered by beach and dune deposits, and its southern shores steered the littoral sand further west. Through time an extensive belt of shoals and emerging islands developed between eastern Dauphin Island and present New Orleans, Louisiana (Figs. 1 and 2), overlying earlier Holocene muds and sands with silty and muddy-sand layers. From the very few reliable radiocarbon dates (near East Ship and under Point Clear Island; Otvos 1978) and from the fact that no subaerial sand deposits (with the possible exception of East Ship Island) have been found on the islands below sea level, the maximum age of the present islands is tentatively set at 3-4,000 years. This period coincides with the start of the relative stabilization of eustatic sea level.

The Holocene sequence under the islands clearly indicates shoal-bar-aggradational island origins. Alternatively, island formation through barrier spit segmentation would seem unrealistic. At no time could the Holocene Mobile Point spit have grown across the deep Mobile Bay entrance channel to link up with Dauphin Island. West of that pass, the littoral sediment supply gradually diminishes along the islands. A need for tidal exchange between the Gulf and the Sound would have always maintained numerous and broad inlets and passes. The origin of the Mississippi-Alabama barrier islands from a 180-200 km long, continuous spit that grew from Dauphin Island with subsequent segmentation seems quite unlikely. In all likelihood, even Petit Bois Island was originally not part of a spit that prograded from eastern Dauphin Island. The two triangular, broad segments of Petit Bois (Fig. 13) probably emerged separately from shoals and were linked up later with Dauphin. Such links were possibly breached and reestablished several times during Late Holocene history.

A third alternative, the theory of mainland dune ridge engulfment and subsequent aggradation with rising sea level (Hoyt 1967), is not supported by the stratigraphic sequence of the Sound or the island-shoal belt. The Mississippi-Alabama nearshore stratigraphy clearly indicates (Otvos 1970a and b) that Hoyt's (1967, p. 1127; 1970, p. 3780) main requirement (the presence of shallow neritic, "open marine" sediments and fauna under lagoonal deposits) for a bar-aggradational barrier island origin is not valid. Due to fresh water dilution, even along certain unbarred mainland coasts, such deposits would not be necessarily expected beneath the brackish deposits of a later established lagoonal system. A change from unbarred to lagoonal/sound setting is nevertheless shown by changes in the biofacies. The facies sequence reflects upward decreasing salinities in several Mississippi Sound Holocene sections (Figs. 5 and 6).

Post-Barrier Developments

With the establishment of the Mississippi Sound and the subsequent encroachment of Mississippi River-St. Bernard subdeltas to the west and southwest (3000-1800 yrs BP; Frazier 1967), salinities have dropped in the general area and westward flow of sand along the islands stopped past Ship Island. The western Sound became much narrower (Otvos 1978, Figs. 16-5, and -6) and the marshland shores prograded steadily. Nearshore Point Clear and Campbell Islands became surrounded by marshland (Fig. 1). The present Pearl River delta and Lake Pontchartrain were established.

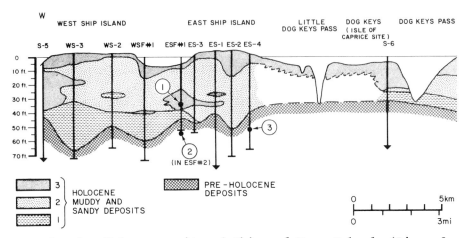

Fig. 8. Holocene units of Ship and Horn Islands (Line of cross-section: Figs. 1 and 2). Symbols: as Fig. 6.

After river flow ceased in the subdeltas, the shorelines also retreated in the south Hancock area. Cat Island, its sand supply from the other islands still cut off by surrounding shoal waters, kept eroding on its eastern end. Subsidence has converted some of the island's progradational, inter-ridge swales into narrow, long embayments. Some of the eroded sediments were shaped by littoral drift into two north-south-trending spits that exhibit storm-dependent, erosional-accretional, cyclic development patterns.

Partial destruction of the St. Bernard subdeltas was accompanied by evolution of the Chandeleur Islands from reworked deltaic Gulf bottom deposits. Judging from the east-west width of Gulf bottom sediments with Mississippi River-type mineralogy (Goldstein 1942) and from the rate of island migration, it is perhaps not unreasonable to assume that the original site of the island chain was some 15-20 km east of the present location.

ISLAND ALTERATION IN THE HISTORICAL RECORD

Erosional Processes

Erosion on the updrift ends of the Mississippi-Alabama barrier islands is a constant, steady phenomenon, but hurricane-surge destruction and segmentation--although infrequent and

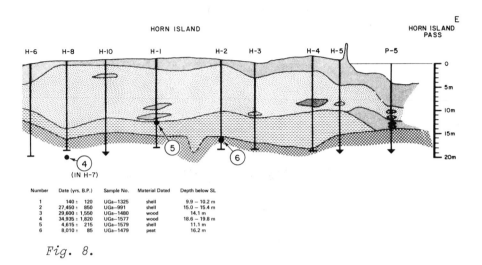

Fig. 8.

episodic--appears to be at least as important a process as littoral drift. Storm segmentation has been recorded numerous
times on the islands; the most important event involved separation of Petit Bois and Dauphin Islands. Early accounts
(Governor Cadillac's 1713 map; Kennedy 1976) and charts (Dé-
lisle 1718; Sr. du Sault 1718-19; Serigny 1719-20; and other
French surveys, partially incorporated in the 1732 d'Anville
map, indicate the existence of a land connection between the
two islands at that time. Unfortunately, subsequent French
(Broutin-Saucier 1743; Bellin 1744 and 1763-64) and British
charts (Jefferys 1755, 1762-63, and his 1775 West India Atlas)
were merely copies of island outlines from earlier maps and
surveys. Therefore, there is no firm proof for how long the
connection lasted. "Half" of the six league (about 35 km) long
Isle Dauphine was washed away during the 1740 hurricane (Gayarré
1903; p. 515). This meant at least temporary breaching of the
island. Bernard Romans' 1773-74 charts, partly based on his
own 1772-73 survey, show a wide gap between Dauphin and Petit
Bois (then, Massacre) Islands, probably not caused by the 1772
hurricane (Otvos, in preparation). Romans (1775) does not
mention island changes in his personal account of that destruc-
tive hurricane and notes, referring on his 1773 charts to the
previous connection between "Dog, Massacre and Dauphine
Islands," clearly imply that the link was cut well before
1772. In addition, Grand Gosier Island, destroyed by the same
storm, is still fully shown on the 1773 Romans chart. The 1776
"Jefferys" (Sayer and Bennett) The American Atlas, and the

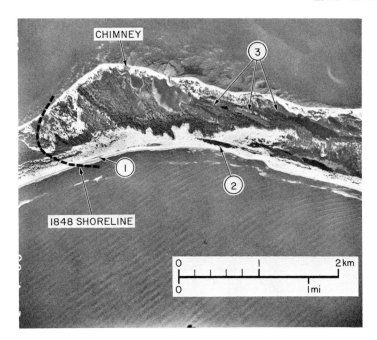

*Fig. 9.   Island growth by spit-attachment, western Horn
Island. 1 - Spit with miniature lagoon behind; 2 - elongated
("cat's-eye") pond, formed by spit-reattachment to island shore;
3 - central island ridges and swales from earlier progradation
island growth stages (USDA aerial photo, 1958).*

1777 "Jefferys" (Faden)   The North American Atlas, as well as
the 1794 (Laurie and Whittle) reprint also show   "Massacre"
and "Dauphin" Islands to be separate. This strongly suggests
that these British maps also relied on pre-1772 data. If the
1740 breach ever healed, it might have been reopened by the
lesser 1759 or 1766 tropical storms.

Comparison of U.S. Coast (later, U.S. Coast and Geodetic)
Survey charts from the 1850s suggests that the reduced island
was segmented by the August 1852 hurricane. Similarly, an 8.5
km wide cut that divided Dauphin Island into two major segments
with some other minor islets, probably resulted from the July,
1916 hurricane (Otvos, in preparation) and not (as has been
stated previously) the 1926 storm. The later hurricane may have
kept the cut open longer. In 1948, the last recorded breach
(Hardin et al., 1976) was minor (0.5 km wide) and healed
quickly.

According to the records, central Ship Island was temporarily cut at least four times during the past 130 years (1852,
1893, 1947, and 1965). As in the case of Dauphin Island, the

Fig. 10. Western tip of Dauphin Island with recurved, progradational spit-swale sequences. (Photo by LCDR Richard N.V. Norat.)

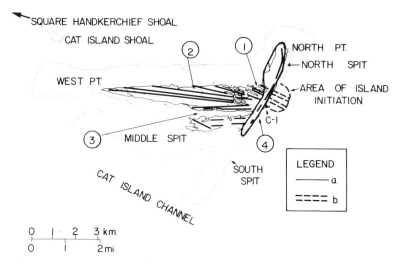

*Fig. 11. Progradational ridge generations, Cat Island.
a - progradational ridges; b - assumed earliest ridge trends
in early island history.*

breaches were across the low, narrow spit-like neck of the
island that offered the least resistance. The 1969 hurricane
cut the island into three parts. Only the western, 3 km wide
"Camille Cut" became permanent. A 3.5 m deep tidal channel developed gradually in this new pass and a broadly-arching ebb-
tidal delta formed seaward of it.

A hurricane cut was probably responsible for the isolation
of a 1.3 km long, triangular islet east of Petit Bois Island
that survived into the 1940s. Another, 0.5 km long small island
off eastern Ship Island (1916-17 survey) and "Massacre" or
"Hurricane Island (18th-early 19th century charts), east of
Horn Island, probably had similar origins (Otvos, in preparation).

Scores of splintered islands formed during each hurricane
that touched the northern Chandeleur Islands. In contrast to
the higher energy Mississippi Sound island shores, the resulting washover fans are permanent features and form integral
parts of the Chandeleurs (Fig. 14).

Numerous examples of total island destruction have been
recorded in this area, and the fragile southern Chandeleurs
have had the most impressive history in this respect. The 1772
destruction of Grand Gosier was noted in the 1778 Gauld chart

(also in Morgan 1977, p. 112). The 1869-1922 USCGS survey
charts (H-999 and H-4223) indicate a 15.2 km wide "hurricane
clearing" in the south Chandeleurs, subsequently reoccupied by
islands. More recently, the combined effects of hurricanes
Betsy (1965) and Camille (1969) reduced a 25 km long island-
studded belt into shoals, with but a few remnant islets between
the northern Chandeleurs and Breton Island. By October 1969,
only a single (probably intertidal) islet existed at the former
site of Curlew Island, with tiny clusters of reemerging bars
north of it (USCGS aerial photos). Grand Gosier Island's remnant was a line of minute, rounded islands that escaped complete destruction.

Episodic hurricane destruction probably played a key role
in the elimination of the 11-12 km long eastern and central
segments of Petit Bois Island between the 1850s and the present. Dog Keys, a cluster of islets between Horn and Ship
Islands have been reduced to shoals, in all likelihood by the
August 1852 hurricane (Otvos, in preparation). A 2.8 km long
barrier island (Dog Island or "Isle of Caprice") emerged before
1907 or 1908 at this site. It was cut in two by the July 1916
hurricane, further reduced during the 1926 hurricane, and
eliminated by 1940. Similar island destruction occurred constantly on the Mobile Bay ebb-tidal delta (Fig. 12), caused not
only by tropical, but also by winter-spring storm activity.

Constructive Processes

The first historical record of island (re)emergence from
shoals was provided by the 1778 Gauld map with the example of
Grand Gosier Island, following the 1772 hurricane. That general
area, as previously noted, has been repeatedly reduced to
shoals. The 1922 USCGS survey chart (H-4223) notes an islet at
the Grand Gosier site, marked "newly emerged since last survey."
USCGS chart 1270 indicates that by 1947 a whole series of
islands, including 8 km long Grand Gosier, has reoccupied the
south Chandeleur shoal belt. After the 1965-69 hurricane destruction, a 9.2 km long island chain emerged by 1974 from
shoals at the Stake-Sand Islands site (Figs. 14 and 15). Most
of Grand Gosier has also reassembled itself by 1977 from one or
two of the tiny remnant islands, but mainly from newly emerging
and merging islets.

As earlier shown, now defunct Dog Island ("Isle of Caprice")
has also aggraded from subtidal shoals between 1853 and 1908.
The appearance and reappearance of small islands is a common
event on the Mobile Bay ebb-tidal delta (Fig. 12). Examples
for island emergence on the eastern delta-flank include the
emergence of Dixie and Coffa Islands (1868-77).

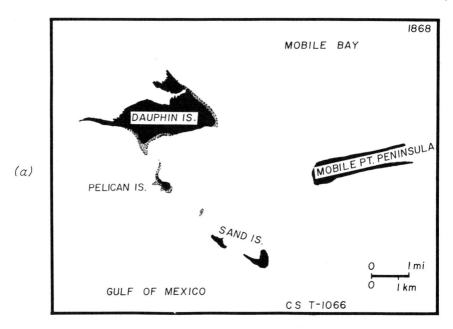

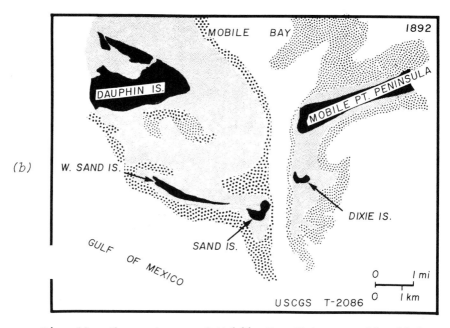

Fig. 12.   Four stages of Mobile Bay Entrance ebb-tidal delta, showing changing configurations of shoals and swash-bar related islands. U.S. Coast Survey and USCGS chart numbers noted.

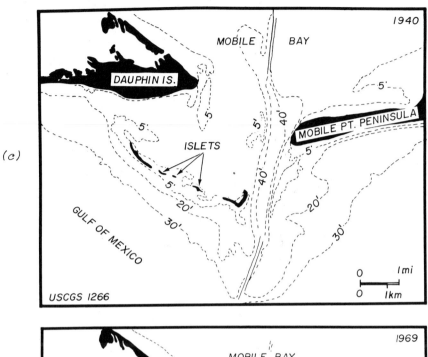

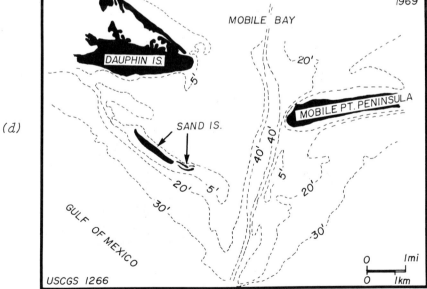

Fig. 12c,d

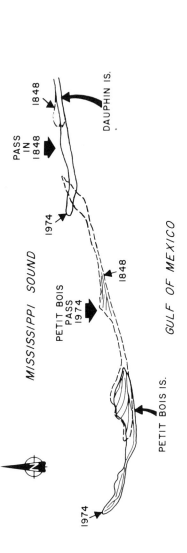

Fig. 13. Migration of Petit Bois Island and Pass, 1848–1974. Fine lines show progradational (originally shore-parallel) dune ridge sets.

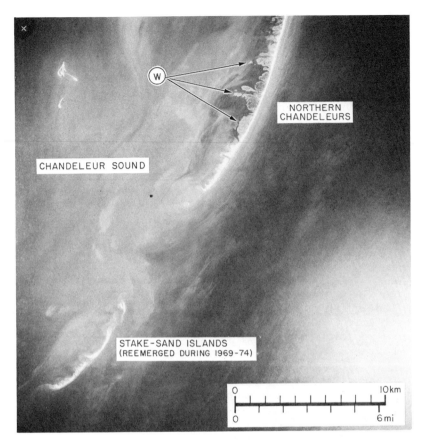

*Fig. 14. Reemerged islands, southern Chandeleur Islands,
Louisiana. W: washover fans. (NASA aerial photo, Mission 293,
October, 1974.)*

Spit growth and spit-reattachment to the beach foreshore is
the major mechanism of island progradation on the Mississippi-
Alabama barrier islands. Seaward widening of the new spit foreshore acts as an additional factor. Elongated, miniature lagoons form behind spits, becoming ponds after a spit reattached
itself to the island (Fig. 9). Occasionally, such a lagoon may
be quite substantial. In southeastern Dauphin Island one large
lagoon served as the French colony's inner harbor in the 18th
century. A golf course presently occupies the site. While most
of the ponds become filled by eolian sand, larger ponds are
preserved even in the island centers as subparallel, elongated
swales (Fig. 9). At the downdrift island tips the spits usually
are recurved, wrapping around the island ends (Fig. 10). This
process accounted for the relatively rapid downdrift island
growth. Landward migration of offshore bars and their welding

onto the islands, typical of mesotidal shore progradation
(Hayes and Kana, 1976), was not noted in this microtidal setting.
    A shore-parallel foredune ridge, that becomes "landlocked"
through the growth of a new seaward spit-lagoon set, is the
most recent addition to a barrier island's progradational dune
ridge sequence. South- and westward-prograding ridge generations with intervening swales are conspicuous on Petit Bois,
West Ship and Cat Islands (Fig. 11) and part of East Ship
Island. The nearly north-south-oriented swales in the center of
East Ship probably are relict hurricane channels, not progradational features.

ISLAND MIGRATION

Mississippi-Alabama Coast

    The accretional-erosional processes that are responsible
for island migration have already been noted. The westward moving islands also overrode and filled a 3-5 m deep tidal channel
in Petit Bois Pass and another one that existed along the west
tip of Horn Island in the 1850s. Presently, prograding Dauphin,
Petit Bois and West Ship Islands are spilling into their associated tidal channels, and maintenance dredging is necessary
to keep the first two open and stationary. Since the 1850s,
Petit Bois Pass (Fig. 13) widened from 2.6 to 8.4 km (1974),
while at least after 1916 (Hardin et al. 1976, p. 80-82) the
position of its tidal channel remained unchanged.
    Historical charts indicate that the islands have maintained
their present distance from the mainland. Although erosion and
accretion occurred both on the Gulf and Sound shores, such
shifts were localized (Table I). For instance, the central,
spit-like, narrow segment of Ship Island moved several hundred
meters northward between 1848 and the second quarter of the
20th century, synchronously with the erosional retreat of the
contiguous south shore of eastern Ship Island. During erosion
of Horn Island's east tip, the south shore of the eastern
island end gradually retreated northward, resulting in little
net change of the island's general attitude. Nearshore current
erosion has allowed storm waves to break closer inshore along
certain north island shores, resulting in considerable shore
retreat. The result is most noticeable at historic Fort Massachusetts, West Ship Island.

TABLE I.   *Erosional-accretional island shore changes*
(maximum values, in km), Mississippi-Alabama, 1848-1974.

Islands	Retreating Shores			Prograding Shores		
	Sound-side	Gulf-side	East	Sound-side	Gulf-side	West
Dauphin	0.2	0.1	none	0.1	0.1	7.4
	(1917-74)	(1917-74)		(1917-74)	(1917-74)	
Petit Bois	0.3	0.5	11.5	none	0.3	4.6
Horn	0.2	1.0	4.3	0.2	0.4	1.3
E. Ship (incl. central parts)	0.3	0.5	1.6	none	none	N/A
W. Ship	0.2	0.1	N/A	0.5	0.2	1.4

Chandeleur Islands

Steady, westward migration is characteristic of these islands (Fig. 15). Treadwell (1955, p. 15) noted exhumed peaty swamp-marsh deposits on their Gulf beaches (also common on the Mississippi barrier islands) which is field evidence of that process. Storm surges move sediments overland (washover fans) and through passes (inlets) between the islands. This "tank-tread"-style cyclic process of landward migration has been described for numerous Atlantic barrier islands (Swift et al. 1971; Pilkey and Field, 1972). Similar to the mid-Atlantic islands, the outside sand supply to the Chandeleurs, as noted earlier, is equally meager. Despite these comparisons, the theoretical possibility that shoal aggradation played a significant role in the original formation of the mid-Atlantic barrier islands has been inexplicably rejected in favor of the other two genetic theories (Swift 1975; Field and Duane, 1976, 1977). In post-storm reemergence from shoals, the new islands occupy positions closer to the mainland shores than their pre-storm locations.

CONCLUSIONS

Origins of the Alabama-Mississippi barrier islands were studied by corehole data. A detailed study and comparison of the microfauna in the core samples was useful in demonstrating that the mainland shore remained unbarred in the early state of the Mid-Late Holocene transgression. Therefore, it appears that the barrier islands did not originate seaward of their present locations. In contrast with the transgressive Atlantic, Texas (?) and other Holocene barriers, the Mississippi barrier islands emerged from shoals practically "in place" and only shifted

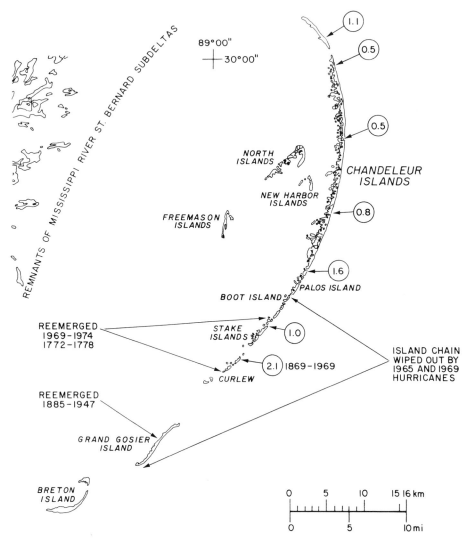

Fig. 15. Migration and reemergence of islands, Chandeleur chain. Migration distances (km) in circles for period between 1885 (USCGS chart #H-1654) and 1969 (#1270). The 1869 positions were based on USCGS chart #H-999.

westward with the littoral drift. Geological and historic records did indicate shoreward migration of the Chandeleur island
chain in Louisiana during the Late Holocene, including the
past 110 years.

Extensive island fragmentation, erosion, progradation and
complete destruction have been documented in the study area.
While we can still only speculate about the specific processes
responsible for the repeated emergence of numerous smaller
islets and one larger island (Grand Gosier) in recent times,
these events mirror the aggradation of numerous other, recently
emerged Gulf Coast islets and of the Late Holocene Mississippi
Coast barrier islands. The eastern end of Dauphin Island,
formed as a high Pleistocene core area, was veneered over by
Late Holocene beach and dune sands, in a manner similar to the
evolution of Deer and Round Islands in the Mississippi Sound,
and others along the southeastern Atlantic Coast.

ACKNOWLEDGMENTS

Wade E. Howat was in charge of both the field and the
sediment-analysis phases of the drilling project. Wayne D. Bock
identified the foraminifer fauna and provided paleoecological
guidelines for their interpretation. Betty Lee Brandau, University of Georgia, provided the radiocarbon age dates. The U.S.
National Park Service provided transportation to four of the
islands involved, and the U.S. Army Corps of Engineers (Mobile
District) gave drill material from two of the coreholes, used
in the present analysis. For help given in the study of historic island changes, I express my thanks to J.A. Bomar, D.K.
Carrington, Fr. J.N. Couture, J. Higginbotham, J.A. Mahe II,
R.W. Stephenson, M.J. Stevens; and several others in six
archives and map collections including the Geography and Map
Division, The Library of Congress.

REFERENCES

Curray, J.R. and Moore, D.G. (1963). Facies delineation by
    acoustic-reflection: northern Gulf of Mexico. *Sedimentology*
    2, no. 2, p. 130-40.
Eleuterius, Ch. K. (1974). Mississippi Superport Study-
    Environmental Assessment. Office of Science and Technology,
    Office of the Governor, State of Mississippi, 248 p.
Eleuterius, Ch. K. (1977). Location of Mississippi Sound oyster
    reefs as related to the bottom waters during 1973-75. Gulf
    Research Reports, v. 6, no. 1, p. 17-24.

Field, M.E. and Duane, D.B. (1976). Post-Pleistocene history
    of the United States inner continental shelf: Significance
    to origin of barrier islands. *Geol. Soc. America Bull. 87,*
    p. 691-702.
Field, M.E. (1977). Post-Pleistocene history of the United
    States inner continental shelf: Discussion. *Geol. Soc.*
    America Bull. 88, p. 735-36.
Frazier, D.E. (1967). Recent deltaic deposits of the Mississippi
    River; their development and chronology. *Gulf Coast Assoc.*
    Geol. Socs. Trans. 17, p. 287-315.
Gayarré, Ch. E.A. (1905). History of Louisiana, v. 1, 540 p.
    (Republished by Pelican Publ. Co., New Orleans, 1965).
Goldstein, A., Jr. (1942). Sedimentary petrologic provinces of
    the northern Gulf of Mexico. *Jour. Sedimentary Petrology 12,*
    p. 737-768.
Hardin, J.D., Sapp, C.D., Emplaincourt, J.L., and Richter, K.E.
    (1976). Shoreline and bathymetric changes in the coastal
    area of Alabama. Geol Survey, Alabama. Inform. Ser. No. 50,
    123 p.
Hayes, M.O. and Kana, T.W., eds. (1976). Terrigenous clastic
    depositional environments. Some modern examples. Techn.
    Report No. 11-CRD, Coastal Res. Div., Dept. Geology, Univ.
    of South Carolina, 315 p.
Hoyt, J.H. (1967). Barrier island formation. *Geol. Soc. America*
    Bull. 78, p. 1125-1136.
Hoyt, J.H. (1970). Development and migration of barrier
    islands, northern Gulf of Mexico: Discussion. *Geol. Soc.*
    America Bull. 81, p. 3779-3782.
Kennedy, J.M. (1976). Dauphin Island, Alabama (French Posses-
    sion, 1699-1763). Coffee Printing Co., Selma, Alabama, 41 p.
Morgan, D.J. (1977). The Mississippi River delta. Legal-
    geomorphologic evaluation of historic shoreline changes.
    Geoscience and Man, v. 16, 196 p. Louisiana State Univ.
Otvos, Ervin G., Jr. (1970a). Development and migration of bar-
    rier islands, northern Gulf of Mexico. *Geol. Soc. America*
    Bull. 81, p. 241-246.
Otvos, Ervin G., Jr. (1970b). Development and migration of
    barrier islands, northern Gulf of Mexico: Reply. *Geol. Soc.*
    America Bull. 81, p. 3783-3788.
Otvos, Ervin G., Jr. (1972). Mississippi Gulf Coast Pleistocene
    beach barriers and the age problem of the Atlantic Coast
    "Pamlico" - "Ingleside" beach ridge system. *Southeastern*
    Geology 22, p. 241-250.
Otvos, Ervin G., Jr. (1975). Inverse beach sand texture -
    coastal energy relationship along the Mississippi Coast
    barrier islands. *Jour. Mississippi Acad. Sciences 19,*
    p. 96-101.

Otvos, Ervin G., Jr. (1976a). Holocene barrier island development over preexisting Pleistocene high ground: Dauphin Island, Alabama. Geol. Soc. America Southeastern Sec. 25th Mtng. Abstr., p. 76.

Otvos, Ervin G., Jr. (1976b). Post-Miocene geological development of the Mississippi-Alabama coastal zone. *Jour. Mississippi Acad. Sciences 21*, p. 101-114.

Otvos, Ervin G., Jr. (1976c). Mississippi Offshore Inventory and Geological Mapping Report. Mississippi Marine Resources Council, 27 p., 9 maps.

Otvos, Ervin G., Jr. (1978). New Orleans-South Hancock Holocene Barrier Trends and Origins of Lake Pontchartrain. *Trans. Gulf Coast Assoc. Geol. Socs. 28*, p. 337-355.

Pilkey, O.H. and Field, M.E. (1972). Offshore transportation of continental shelf sediments: Atlantic southeastern United States. *In* "Shelf Sediment Transport" (D.J. Swift, D.B. Duane, and O.H. Pilkey, eds.), p. 429-446. Dowden, Hutchinson and Ross, Inc., Stroudsburg, Penna.

Romans, Bernard (1775). Concise natural history of East and West Florida, 291 p. (Reprinted by Pelican Publ. Co., New Orleans, La., 1961).

Swift, D.J.P. (1975). Barrier island genesis: Evidence from the central Atlantic shelf. *Sed. Geology 14*, p. 1-43.

Treadwell, R.C. (1955). Sedimentology and ecology of southeast coastal Louisiana. La. State Univ. Techn. Rprt., No. 6 (ONR), Baton Rouge, 78 p.

U.S. Army Engineer Division (1971). National Shoreline Study - Regional Inventory Report. Corps of Engineers, Atlanta, Georgia.

Waller, T.H. and Malbrough, L.P. (1976). Temporal changes in the offshore islands of Mississippi. Water Res. Inst., Mississippi State Univ., 109 p.

SUBJECT INDEX